T0389727

Travel Writing and Cultural Transfer

FILLM Studies in Languages and Literatures
ISSN 2213-428X
For an overview of all books published in this series, please see
benjamins.com/catalog/fillm

Volume 20

Travel Writing and Cultural Transfer
Edited by Petra Broomans and Jeanette den Toonder

FILLM Studies in Languages and Literatures

Travel Writing and Cultural Transfer

Edited by

Petra Broomans
Jeanette den Toonder
University of Groningen

John Benjamins Publishing Company
Amsterdam / Philadelphia

The paper used in this publication meets the minimum requirements of the American National Standard for Information Sciences – Permanence of Paper for Printed Library Materials, ANSI Z39.48-1984.

DOI 10.1075/fillm.20

Cataloging-in-Publication Data available from Library of Congress:
LCCN 2024031284 (PRINT) / 2024031285 (E-BOOK)

ISBN 978 90 272 1587 1 (HB)
ISBN 978 90 272 4654 7 (E-BOOK)

John Benjamins Publishing Company · https://benjamins.com

Table of contents

Series editor's preface

The Fédération Internationale des Langues et Littératures Modernes (FILLM), founded in 1928, is UNESCO's ceiling organization for scholarship in the field of languages and literatures. The Federation's main aim is to encourage linguists and literary scholars from all over the world to engage in dialogue. During the twentieth century, linguistic and literary studies became steadily more professional and specialized, a development which significantly raised the overall standard of research, but which also tended to divide scholars into many separate and often smallish groupings between which communication was rather sporadic. Over the years this became something of a handicap. New ideas and findings were often slow to cross-fertilize. Given the rapidly globalizing world of the early twenty-first century, the relative lack of contact between scholars in different subject-areas became a more glaring anomaly than ever.

In response, FILLM decided to set up its own book series, originally edited by Roger D. Sell. Responsibility for the series rests with an editor, an Editorial Board, and an Advisory Board drawn from the many constitutive organizations of the Federation. Books appearing under the label of FILLM Studies deal with languages and literatures around the world, explore specific cases in order to raise broader problems, and are written in an English that will be immediately understandable and attractive to any likely reader. Every book in the series will present original findings, including new theoretical, methodological and pedagogical developments. We encourage book proposals from authors whose work fits this model.

Haun Saussy

Author biographies

Nina Brandau obtained after her Bachelor studies in Mediaculture and Scandinavian Studies at the University of Cologne an Erasmus Mundus Master of Arts in "Euroculture: Society, Politics and Culture in a Global Context" at the universities of Groningen and Göttingen. In her Master thesis she focussed on cultural transfers between Germany and Scandinavia with a special interest in Max Reinhardt's and Deutsches Theater's activities in Sweden. Currently she works in the research project "Smart Schools", initiated by Helmut Schmidt University Hamburg und KU Leuven. Her PhD project focuses on participatory and critical approaches to digital education design.

Petra Broomans is associate professor e.m. with ius promovendi at Groningen University and visiting professor at Ghent University between 2011 and 2022. In January 2020 she was appointed Doctor honoris causa by Uppsala University's Faculty of Arts. She is the initiator and coordinator of the Dutch translators' dictionary: https://www.vertalerslexicon.nl/. She has published extensively on cultural transfer, Scandinavian literature and women's literature. Her research interests include cultural transfer, world literature, meta-literary history and minority literature. She is series editor of the *Studies on Cultural Transfer and Transmission (CTaT).*
See for an overview of her works: https://www.petrabroomans.net

Andreas Hedberg is associate professor of literature at Uppsala University, Sweden. Among his research interests are publishing studies, processes of literary canonization and the critique of modernity in 19th and 20th century literature. In his most recent projects, he has focused on world literature, especially cultural transfer and the sociology of translation. Hedberg's latest publications include *Northern Crossings: Translation, Circulation and the Literary Semi-periphery* (Bloomsbury, 2022) and "The Othering of Others: Domestication and Foreignization in the Reception of Swedish Literature on the French Book Market 1945–2018" (in *Multilingualität und Mehr-Sprachlichkeit in der Gegenwartsliteratur*, 2019).

Eduardo Gallegos Krause is a lecturer of semiotics, discourse analysis, and communication theory at the Department of Languages, Literature, and Communication at *Universidad de La Frontera* (Chile). His doctoral research focused on the link between European travel accounts and the emergence of journalism in Chile in the 19th century. Besides, he is interested in the influence of travelogues in cultural formations and the representation of identity and otherness. He has completed research stays at the University of Groningen (Netherlands) and the Université Rennes 2

(France). In 2023, he was selected visiting professor at the International Chair of Humanities and Social Science at the Université Rennes 2.

Kirsten Sandrock teaches at the University of Würzburg, Germany, where she holds the Chair of English Literature and Culture. Her research ranges from early modern to contemporary literature, with cross-period interest in transnational literatures, travel writing, gender, Scottish studies, and Shakespeare. Kirsten is author of *Scottish Colonial Literature: Writing the Atlantic, 1603–1707*, published with Edinburgh University Press in 2021. She is currently vice president of the German Shakespeare Society and has a continuing interest in Scottish travel writing.

Marja van Tilburg has been affiliated to the Department of History of the University of Groningen, Netherlands, since 1986. She contributed to the interdisciplinary Minor Gender Studies and served the board of the Centre for Gender Studies of the Faculty of Arts. Her thesis *Hoe hoorde het? Seksualiteit en partnerkeuze in de Nederlandse adviesliteratuur 1780–1890* (Amsterdam: Het Spinhuis, 1998) explores the diffusion of Enlightenment pedagogy in conduct books with regard to sexuality and gender. Following up on this research, she has explored the reception of Enlightenment theory on cultural diversity in travel literature of the turn of the 1800.

Jeanette den Toonder is assistant professor at the Department of European Languages and Cultures at the University of Groningen, and Director of the Centre for Canadian Studies. Her research interests include questions of identity, autobiography, journey and space in the contemporary French and francophone novel. She is particularly interested in migrant and minority writing and the female voice and has published on "écriture migrante" and First Nations literature in Québec. In her current research project she examines traumas of migration, linguistic interference and migratory feminism in contemporary novels published by female authors from Iranian descent. For an overview of publications: https://www.rug.nl/staff/j.m.l.den.toonder/

Weishi Yuan is research assistant and advisor for international affairs at Rheinische Friedrich-Wilhelms-Universität Bonn. He holds a BA in English Studies from Xiamen University and a MA in Euroculture from Rijksuniversiteit Groningen and Georg-August-Universität Göttingen. His research interests include cultural transfer, comparative literature and intellectual history. In 2021 he obtained his PhD (*magna cum laude*) from Georg-August-Universität Göttingen. His dissertation focuses on Albert Schweitzer's interpretation of Chinese philosophy in the framework of the ethical philosophy of Reverence for life. In particular, it investigates Schweitzer's reception of European sinologists such as Richard Wilhelm within a quarter of a century.

Introduction

Travel writing and cultural transfer

Petra Broomans & Jeanette den Toonder
University of Groningen

This book is about the association of travel and cultural transfer. To travel means to cross borders and to explore new territories by horse, coach, boat, train or aeroplane, and very often also on foot, following in the footsteps of famous people, as did Richard Holmes. In his marvellous book, *Footsteps. Adventures of a Romantic Biographer* (1985), Holmes reconstructs the journeys of Robert Louis Stevenson through the Cévennes, Mary Wollstonecraft in Paris, Percy Bysshe Shelley in Italy and Gérard de Nerval in France, among others. Both Stevenson and Wollstonecraft are protagonists in chapters in this volume, which is concerned with theories of travel writing and cultural transfer, with case studies exploring the links between them. *Footsteps* is a fine example of travel writing, with Holmes not only reconstructing the journeys by consulting archives, letters and travelogues, but also making the journeys himself, attempting to travel in the same way as his predecessors. Furthermore, he reflects on his own experiences and emotions while travelling and visiting the places where his protagonists have been.

Thus, Holmes' *Footsteps* is biographical as well as autobiographical, which foregrounds one of the key issues in both travel writing and cultural transfer studies: the role of the writer and the transmitter. Holmes describes the lives of the travellers, the times in which they lived and the places and countries they visited. Maps are included and the reader can follow Shelley, De Nerval and the others step by step. Holmes transmits knowledge about circumstances in the eighteenth and nineteenth centuries, especially the cultures that his travellers encountered and wrote about in their letters and diaries. We argue that *Footsteps* can be regarded as a travel writing narrative that links factual to fictional worlds and, in so doing, emphasises the significance of both to processes of cultural transmission. It includes real journeys, with Holmes literally wandering in the footsteps of travellers in the past, and imagined travels through his reconstruction of the journeys. This presentation of travel as both factual and fictional is one of the many forms travel writers have used to illustrate the fluidity of the genre, with the authors and narrative both moving between different worlds. This is one of the

https://doi.org/10.1075/fillm.20.intro

reasons that processes of cultural transmission in this genre are so manifold and powerful. We will further reflect on the generic fluidity of travel writing in relation to the forms presented in this volume towards the end of the introduction.

New means of transportation that developed in the first decade of the twentieth century allowed for more extensive travel. The expansion of train networks in Europe was extremely significant, not only for transporting people and goods but also for the transfer of ideas, opinions and aesthetic movements, as Orlando Figes demonstrates in his study, *The Europeans: Three Lives and the Making of a Cosmopolitan Culture* (2019). Figes approaches Europe as a "space of cultural transfers, translations and exchanges crossing national boundaries."[1] To a certain extent, similar developments took place in other parts of the world, where cultural transfer is not infrequently linked to issues of power, colonialism and politics, as various chapters in this volume illustrate. The development of the railways, the telegraph and the open market economy thus shaped both Europe and the world, and although technological advance has generally moved from the physical to the digital world, we still live in a time of increasing global connectivity. In the early twentieth century, the translation trade also became more international, although this was also accompanied by an increase in nationalist feelings. This illustrates some of the ambiguities underlying cultural transfer processes,[2] with anti-cosmopolitanism one of the reasons for the outbreak of the First World War.

Travel by train meant not only border crossing, but also reading and writing on the train, while postcards and letters were sent from remote places, with observations during the journey recorded on the train or at new destinations. The genre of travel writing is thus closely connected to mobility and space, to connecting and transferring, as is demonstrated by contemporary research in the field (e.g., Das, Nandini and Youngs 2019, Schaff 2020) and by the contributions in this volume. Thus, it is arguably an important vehicle for cultural transfer.

Having developed from the field of comparative literature, cultural transfer studies initially defined cultural transfer as the process of mediating and/or translating a literary, cultural or historical text from one linguistic area to another. However, travel writing allows us to broaden our understanding of the traditional process of cultural transfer, not only but especially because of its complex relationship to factual and fictional worlds.[3] The contributions in this volume inves-

1. Orlando Figes, *The Europeans: Three Lives and the Making of a Cosmopolitan Culture* (New York: Metropolitan Books, 2019), 4.

2. As Figes remarks, there was growing opposition to cosmopolitanism in Europe after the 1870s. "The influx of foreign books led in many countries to protests by those who feared that it would undermine the distinctive nature of their domestic literatures." Figes, *The Europeans*, 475–6.

tigate the transmission of other cultures, ideas and ideologies to the writer's own cultural sphere and consider how the processes of cultural transfer interact with the forms and functions of travel writing.

The travel writer not only takes on various roles, such as explorer, missionary, reporter, entrepreneur, Grand Tour traveller, investigator, travelling artist and writer, but also acts as a cultural transmitter. In the nineteenth century, the tourist traveller can be added to the list, while after the Second World War, the mass tourism industry played an enormously important role in the economics of travelling, with big players such as Booking.com and cheap airlines such as Ryanair. These swift journeys generally do not generate fascinating travelogues, but they do inspire travellers to publish well-written blogs on internet forums and essays in literary journals. Thus, they introduce a new dimension to travel writing as a genre and add different forms of cultural transfer to the traditional media and narrative practices.

The travellers presented in this volume made long journeys, with some never returning to their home countries, and they offer a variety of insights into and perspectives on the travelling cultural transmitter. We selected individual travellers, known and famous or forgotten and less famous. They all wrote seminal travel accounts in which they experience various encounters with or within other cultures, or who can be regarded as practicing cultural transfer and transmission by travelling and working in other cultures, and thus finding themselves in a new culture and environment.

In order to accentuate how the contributions to the volume are connected, this introduction presents them from various perspectives. After a reflection on the development of the genre of travel writing in relation to the contributions in this volume, we will further examine how travel writers manifest themselves as cultural transmitters in the discussed travel accounts. The final section of the introduction will assemble the key concepts developed across the various chapters, and thus connect them on a more abstract level, establishing a framework for the process of cultural transfer in travel writing. The chapters in this volume are arranged chronologically in order to display the development of travelling, the way of travelling and the way travellers perceived the cultures they visited.

3. See, for the basic definition, Petra Broomans, "Preface," in *From Darwin to Weil. Women as Transmitters of* Ideas, ed. Petra Broomans (Groningen: Barkhuis, 2009), vii.

Generic fluidity

Travel writers have many forms at their disposal to establish contact through writing. As the example of Holmes' *Footsteps* demonstrates, the factual, the biographical, the autobiographical and the fictional can all be intertwined in the travel writing process. It is therefore not surprising that the genre of travel writing has been discussed extensively in literary studies; in particular, since the 1980s, when it became a new field of study. In his study, *Travel Literature and the Evolution of the Novel* (1983), Percy G. Adams, for example, regarded travel literature as an important genre for the development of the novel. The most urgent problem, however, remains how to define the genre. This is also pointed out by Barbara Schaff in her *Handbook of British Travel Writing* (2020): "[...] the historical specificities, narrative diversity and complex formal aspects of travel writing have continued to present scholars with classificatory challenges. Many have commented on how difficult a generic demarcation turns out to be."[4] Our focus here is on narrative diversity rather than on the actual demarcation of the genre, because it is precisely its hybridity, its fluidity, that allows for the many ways in which cultural transfer takes place.

In early modern culture, the travelogue developed as a genre linked to both internal and external expansion. It frequently included a chronological overview of events, a description of the local society and a list of useful words in an appendix, as is demonstrated in this volume by Marja van Tilburg, who presents the example of Bougainville's travelogue. Some varieties in the development of the travelogue genre can be observed, such as compendia and imaginary travel accounts. Another development seen in Bougainville's travelogue, as noted by Van Tilburg, is its ethnographic character. Writing about their journeys, travel writers often aim to provide information on different cultures and societies, and not only to transmit their experiences, personal emotions and impressions. As a mixture of "faction" and fiction, these texts, such as Mary Wollstonecraft's *Letters written during a short residence in Sweden, Norway and Denmark* (1796), illustrate that travel writing is a hybrid genre.

When relating travel writing to the development of the novel, it can be argued that the travel account is subjective rather than objective.[5] In her contribution to this volume, Petra Broomans contends that Wollstonecraft's *Letters* can be read in this light and that they also fit Schaff's description of travel writing as a mercurial category "involving, and sometimes combining, elements of autobiogra-

4. Barbara Schaff, *Handbook of British Travel Writing* (Berlin/Boston: De Gruyter, 2020), 1.
5. See Percy G. Adams, *Travel Literature and the Evolution of the Novel* (Lexington: The University Press of Kentucky, 1983).

phy, memoir, diary, letters, journalistic reportage, fiction, poetry, satire, or travel guide."[6] According to Christoph Bode, during the Romantic period, the self develops "towards a highly subjective, self-reflective discourse of the travelling persona."[7] From this perspective, Mary Wollstonecraft's travel account also reflects a "turn towards self-reflectivity."[8]

It is, however, important to be aware of the context and the function of the travel account. It can also be considered as an informative text, when, for example, instructions are given to the traveller before the journey. If the travel is for pleasure, readers would expect another kind of text. This expectation is determined by what Lidström calls a "genre contract."[9] Lidström regards the category of the "periploi," or guides for travellers, as examples of functional travel texts.[10] In applying Lejeune's definition of the genre of autobiography to travel writing with respect to the narrator, Lidström considers the author, narrator and protagonist as one and same person.[11] Thus, travel literature is an autobiographical genre that enables the travel writer to choose between roles.

The basic role of the travel writer is, of course, to be "the traveller," but there are several additional roles. It is here that Lidström uses the term "persona."[12] Examples of persona in travel writing include the adventurer, the reporter, "our man on the spot,"[13] scholars and aesthetes. The persona in travel writing often uses paratexts, which we also observe in Wollstonecraft's *Letters*, in which she constructs a new form of travel writing from a gendered perspective. This connects the *Letters* with more recent examples of travel narrative, such as the story of a young woman in exile in *Désorientale* (2016), written by Franco-Iranian author Négar Djavadi and discussed in this volume by Jeanette den Toonder. Den Toonder argues that the narrator's encounter with different cultural values enables her to reflect on her body and sexuality and to situate herself in relation to others. This experience is reflected in the travel narrative, where the individual story of encounters with others intertwines with stories of family members and other exiled figures. In *Désorientale*, gender, travel account and migration shape a genre

---

6. Schaff, *Handbook*, 1.

7. Cited in Kirsten Sandrock, "Discourses of Travel Writing," in *Handbook of British Travel Writing*, ed. Barbara Schaff (Berlin/Boston: De Gruyter, 2020), 38–9.

8. Sandrock, "Discourses," 39.

9. Carina Lidström, *Berättare på resa. Svenska resenärers reseberättelser 1667–1829* (Stockholm: Carlson, 2015), 23.

10. Lidström, *Berättare*, 23.

11. Lidström, *Berättare*, 31.

12. Lidström, *Berättare*, 35.

13. Lidström, *Berättare*, 74.

in which the female subject describes the exiled self on the road, or roaming the world as a nomad, in connection with others.

The figure of the nomad and the genre of life writing inspire other contributions in this volume. Examining Richard Wilhelm's travel writings, which were compiled in *The Soul of China* (1925), Weishi Yuan observes the personal way in which Wilhelm reports on the remarkable encounters he had during his 25 years of travel, as if noting them in a diary. As Andreas Hedberg demonstrates, the nomadic experience of the traveller in Harry Martinson's accounts is very complex, changing from the perception of nomadic existence as an ideal to the conception of the nomad as a traveller disinterested in the world, longing not for mobility but for attachment; for an absolute space rather than uprootedness.

Not only written texts, but also other artistic forms of expression, such as images and theatre performances, inform the development of travel writing. The analysis of the travels of Max Reinhardt and Alexander Moissi by Nina Brandau in this volume broadens the scope of the genre, including the exchange of ideas within the context of European theatre. Performances develop into collective events which shape intercultural connections. The reproduction of images in travel writing strengthens or weakens positions expressed in the narrative, as is the case in Nordenskjöld's notions of self and other. In his contribution to this volume, Eduardo Gallegos Krause, therefore, refers to such ambiguities in travel writing, as reflected in the relation between text and image.

The blending of genres thus shapes the development of travel writing, which becomes a crossroads of narrative forms and the perfect site for encounter and cultural transfer. This dynamic character is illustrated by the opposition of fixity and movement observed by Kirsten Sandrock when focusing on the concepts of space and time in Robert Louis Stevenson's Pacific writings. As a physical object, the literary work is partially fixed in space and time, while the travelogue is a genre that is particularly based on and closely connected to mobility and (ex)change. This paradox between the insecure processes of cultural transfer and the solid material form in which they are recorded forms part of Stevenson's aesthetic construction of his work. He presents these ambiguities through the mediation of different chronologies and different worldviews.

Writing about encounters also means addressing conflict. Writing about the possibility of not understanding the other sometimes involves the expression of prevailing ideas. Narrative hybridity expresses this longing as well as this sense of ambiguity and insecurity of the travel writer. The fluidity of the genre offers the possibility of engaging with these contrasting feelings in a productive way, allowing for meaningful cultural transfers to occur.

The travel writer as cultural transmitter

The contributions in this volume are organised in chronological order with each chapter focusing on a particular case study that illustrates the specificities of travel writing in a given period. With regard to the spatial realms examined, the various chapters explore contact between European travel writers and other parts of the world – notably the Pacific, China and Antarctica – as well as exchanges between or within various European countries – in particular Great Britain, Sweden and Germany. The travel writers in this volume are European and either leave the continent or travel through it, with the exception of the final chapter, which focuses on the phenomenon of exile. In this case, the travel writer, who is Iranian in origin, moves to Europe. It is necessary to note these spatial dynamics, because they are linked – as are many processes of cultural transfer – to questions of power, politics and cultural expansion.

As a result of colonial expansionism from the sixteenth century onwards, the travelogues of the historical traveller-explorers frequently present their encounter with overseas territories and cultures, connecting the "old" and the "new" worlds that Europeans explored and often exploited. The case study by Marja van Tilburg examined within this framework is Admiral de Bougainville's report, *Voyage autour du monde* (1771), and the philosopher Diderot's comments on it in *Supplément au voyage de Bougainville* (1776). The texts illustrate the Enlightenment's interest in non-European practices as a basis from which to critique and advance contemporary European society. With regard to the traveller-explorer's capacity as cultural transmitter, Bougainville's report is extremely interesting. Van Tilburg's analysis of the process of cross-cultural encounters in the framework of Enlightenment thinking connects the travelogue and the critical commentary on it. The explorer tries to render Tahitian culture comprehensible to European readers and the *philosophe* rewrites Bougainville's descriptions in accordance with the Enlightenment discourse. The engagement with Tahitian practices, Van Tilburg argues, thus results in a cultural transmission by means of critical thinking, supporting Enlightenment ideas of social improvement and the better functioning of French society.

Within Europe towards the end of the eighteenth century, travel permits a comparison of and reflection on different national characters. Mary Wollstonecraft's *Letters* not only offers the possibility to study the mediation of national characters, ethnotypes, political and societal ideas, but also illustrates the role of women as innovators in the public sphere. Combining commerce, facts, emotions and a poetic veil, the letters are particularly interesting with regard to the role of female travel writers. As Petra Broomans demonstrates, Wollstonecraft combines her ideas on the national character of the Scandinavian people she meets with the expression of her persona, the self, the woman grieving lost love, thus introducing a modern subjective perspective. This also characterises her role as a cultural transmitter, connecting her scholarly attitude with her personal experience.

Towards the end of the nineteenth century, the travel writer becomes increasingly aware of a movement of mutual exchange, in which cultures are not only observed and interpreted from a particular perspective but can also be connected. The varied cultural encounters in Stevenson's *A Footnote to History* (1892) and *In the South Seas* (1896) exemplify how the traveller-narrator functions as an intermediary figure. As Kirsten Sandrock argues, Stevenson's narrative use of an insider-outsider perspective creates a connection between the traveller-narrator, the Pacific cultures written about and the Western audiences addressed. In this way, Stevenson provides an example of how travel writing may attempt to overcome temporal and geographical distances. Being aware of the critique of Western colonialism and the way in which it has imposed Western conceptions around the world, the travel writer illustrates the importance of mediating cultural knowledges.

However, moving between two worlds, the traveller is not always capable of leaving behind a European perspective, even when colonial violence is explicitly condemned. This is the case in the travelogue "La terre de feu" (1902), analysed by Eduardo Gallegos Krause, where the Swedish geologist, geographer and explorer Otto Nordenskjöld criticises the colonisation of the Indigenous people of Patagonia, while still considering them to be "the others." Focusing on Nordenskjöld's report on his experiences in Patagonia, Gallegos Krause argues that, while the Swedish traveller-explorer indeed mediates between Indigenous and European cultures in his travel writing, his focus on the notion of alterity reiterates traditional Western stereotypes at the same time. This example of the encounter with the Global South illustrates the ambiguous attitude of the European traveller-explorer and the difficulty of establishing relationships of equality between cultural communities.

It is not only explorers and authors who have been important contributors to the exchange of cultural and artistic ideas, but also representatives of other cultural fields, such as directors, actors and translators. Nina Brandau's study of Max Reinhardt's and Alexander Moissi's guest performances, bringing the German theatre company to Stockholm between 1915 and 1921, demonstrates how the development of modern theatre has benefited from artistic mobility. The concept of cultural transfer in its traditional sense thus starts to expand during the First World War, including performance and focusing on the synthesis of different theatrical elements. The mobility of artists and directors allowed for guest performances that were mutually beneficial for both cultural areas, and which were facilitated by social, political, aesthetic and infrastructural factors in both countries.

This development also increasingly concerned translation and translators, who can introduce foreign cultures to a wide audience. The traveller-translator

Richard Wilhelm, who moved to China as a missionary but ultimately became an advocate of Confucianism, published his translation of the *I Ching* in German in 1924. As Weishi Yuan observes, Wilhelm searched for similarities between Eastern and Western perceptions of the spiritual world and his work inspired many European intellectuals and artists. As a translator, Wilhelm mediated between Western perceptions of the soul and traditional Chinese philosophy. His cultural translation of this philosophy was further developed in his travelogue, as Wilhelm also adopted the role of traveller-narrator. This allowed him to connect Chinese and European perspectives in various ways, presenting the guiding forces of the soul of China, Taoism and Confucianism to a broad Western audience.

After the First World War, the development of modern transportation and technology offered new forms of travel to cosmopolitan writers who considered themselves nomads of the modern world. This nomadism was not necessarily linked to their membership of a larger community that travelled for work but to individualism and a twentieth-century indulgence in travel for travel's sake. In his contribution, Andreas Hedberg reveals how the Swedish writer Harry Martinson initially shared this ideal of a nomadic existence, but witnessing the emergence of totalitarian regimes in the 1930s, he turns to dystopian narratives in which the journey becomes interstellar. The shifting attitudes in Martinson's travel writing demonstrate that cultural transmission can take different forms. Initially, Martinson's ideal of the traveller was the "geosopher," for whom travelling is a basic need. The geosopher steps beyond the limitations of the nation and its culture in order to experience the world. From this primary global perspective, Martinson's travel writing returns to the microcosm of the Swedish countryside and ends with a dystopic imaginary in his space epic *Aniara* (1956), in which travelling is essentially a painful experience. Martinson engages with his audience in different ways and invites them to reflect on contemporary challenges, as is the case in many recent travel narratives.

Developing out of much postcolonial and migrant writing of the twentieth century, the first decades of the twenty-first century, characterised by exile and displacement, witness a shift from the privileged European travel writer to the migrant writer who is forced to leave home. Using the example of Djavadi's *Désorientale*, Jeanette den Toonder demonstrates how travel writing enables the exile to come to terms with feelings of disorientation and uprootedness, and to share these experiences with a Western audience. The exiled writer is thus particularly well placed to act as a cultural transmitter. The disorienting experience of exile includes the strangeness of the new language and culture and, in particular, cultural codes of family, gender, sexuality and the body. By transmitting her personal story of cultural and bodily estrangement, the female narrator of *Désorientale* creates a space of encounter with the reader where she can also connect her Iranian past and her French present.

The reflections on the development of the travel writer as cultural transmitter in the contributions to this volume, illustrate how cultural transmission is concerned with an encounter with habits and rituals from other cultures and comparing them with the transmitter's own contemporary vision and ideals, as was also the basis of comparative literary studies developed at the end of the nineteenth century. This comparative perspective allows, for example, for reflection on different national characters in the European framework. After having focused on comparison and alignment with their own culture, the travel writer increasingly becomes aware of a mutual exchange between cultures. The role of the travel writer as cultural mobiliser is developed from the end of the nineteenth century onwards, even if this mediation between cultures may be ambiguous, given that it is frequently influenced by the practice of othering. At the same time, successful attempts to build bridges between cultures indeed bring distant practices, beliefs and worldviews closer together.

In the course of the twentieth century, encounter, contact and exchange have become key words in the study of cultural transmission, particularly in the artistic and literary fields. This also involves engagement with contemporary artistic and societal developments, resulting in artistic renewal, a renaissance of literary genres such as the dystopic narrative and the emergence of new genres, notably migrant literature. While the travel writer does not adopt a static role as cultural transmitter, and sometimes experiences travel as alienating and distressing, the mediation between cultures is a fundamental feature, as is the effort to engage various audiences in exciting and sometimes difficult encounters. An important condition of successful transmission is the personal engagement of the travel writer, a feature that runs through the various examples examined in this volume.

The contributions to this volume highlight the large variety of travel writers, who, despite their different backgrounds and the various timeframes in which they wrote or published their work, all contributed to establishing contact between cultures, thus acting as mediators between these cultures.

Connecting key concepts

As observed above, the chronological order points to some developments in travel writing and the travel writer as cultural transmitter over time. When looking further into the various chapters, several compelling conceptual observations that connect the case studies at an abstract level can be made. Travel and cultural transfer are evidently the two main connecting concepts. One of the particular aspects of travel in relation to cultural transfer developed in various chapters in this volume is the travelling of ideas. The transfer of artistic ideas is the centre of attention in Brandau's chapter, where she observes, for example, that multidimensional networks in which objects and ideas circulate play a crucial role in the process of transfer. This includes networks between individual people, networks connecting different artistic levels of theatre and also broader institutional networks. As Broomans and Den Toonder illustrate, travelling to unknown countries and regions awakens new ideas in the travel writer. Moving through the Scandinavian countries, Mary Wollstonecraft develops her ideas on political and societal issues, conveying these to her readers from a personal perspective. Having fled to France, the narrator in Djavadi's novel experiences Western values at first hand, while at the same time sharing insightful ideas and personal stories about Iran that counter Western stereotypical views of the Middle East with which she is confronted every day. As a result, personal as well as cultural values and ideas are critically assessed and transmitted in narrative. The travelling of ideas also relates to intellectual movements, as is demonstrated by Van Tilburg when focusing on Tahitian sexual practices described in Bougainville's report. The passages particularly relating to the role of women are rewritten by Diderot in order to develop ideas on what the Enlightenment discourse presents as the "natural woman."

In order for these travelling ideas to substantially contribute to cultural transfer, contact zones have to be employed or created. This prominent concept of the contact zone takes shape in various ways.[14] Brandau argues that theatres and performances are the contact zones where artistic ideas are mediated. Several spaces are crucial to this process: from the stage, where the contact between actors takes place, to the institutional setting of the theatre, offering possibilities for staging in a new context, and then to the metropolises where theatre institutions are located. The geographical dimension of the contact zone is also evident in Stevenson's works, but at the same time, the analysis of his travel writing highlights the significance of time-based contact zones, where past, present and future meet. As a result of geographical and temporal intricacies, Sandrock demonstrates that cultural contact zones present complex chronotopical entanglements. In his analysis of different forms of cultural transfer in Richard Wilhelm's works, Yuan also refers to complex entanglements, but in relation to different mind-sets at the beginning of the twentieth century in Germany and China. By looking specifically at the two-way transfer of the concept of the soul, Yuan argues that the contact zone adopts a mystical rather than a physical dimension, bringing together Western esoteric traditions developed after the First World War and the ancient Chinese philosophy of Confucianism.

---

14. Mary Louise Pratt, *Imperial Eyes: Travel Writing and Transculturation* (London: Routledge, 1992). A glossary of key concepts for travel writing studies has been published in 2019 by Charles Forsdick, Zoë Kinsley and Kathryn Walchester.

At times, the contact zone also emphasises the unequal power relations between cultures, where the dominating culture imposes its values on the other. One example of this is the loss of autonomy of the Indigenous peoples of Samoa, as recognised by Stevenson. The author tries to resist linear concepts of time imposed by Western colonial power by conjuring up intermediate figures that are not tied to time, such as ghosts. In order to illustrate how well-intended attempts at cultural transfer are still informed by asymmetrical relations of domination, Gallego Krause proposes the term "weak border zones." He demonstrates this phenomenon by examining the discourse that Nordenskjöld adopts when describing images of the Indigenous people.

The discourse, or narrative itself, can also function as a contact zone. One example of this is *Désorientale*, where French and Iranian cultures are seemingly irreconcilable until the narrator expresses her traumatic experience of exile in a personal narrative. The two cultures that initially caused the narrator's disorientation can ultimately be connected by means of this narrative. As these examples demonstrate, cultural contact zones are negotiated in a variety of ways and establish important loci in which travel writers assume their various roles as cultural transmitters.

This discussion of the contact zone also points to the role of space and time in the process of cultural transfer. Cultures are not only separated geographically but also have different temporalities. The concept of time has generally been overlooked in the study of travel writing, and the chapters by Sandrock, Gallegos Krause and Hedberg thus fill an important gap by focusing on diverse cultural approaches to both time and space. Discussing Stevenson's travel works, Sandrock illustrates that cultural transfer is informed by both temporal and topographical exchanges, and the different time zones that cultures inhabit are necessarily involved in the process of mediation. Moreover, in order for cultural mediation to escape the normalisation of difference, cultural time should be considered in the analyses of moments of contact, as Gallegos Krause argues. In the twentieth century, however, as a result of technological innovations, spatial and temporal distances have been increasingly elided. Hedberg discusses how the struggle with time-space compression is mirrored in Martinson's dystopic narrative, but also offers perspectives on ways to integrate the global and the local. Following Doreen Massey's notion of a "global sense of place,"[15] which consists of networks of social relations, a reconnection of the spatial with the temporal is nevertheless conceivable, and it is essential for the liberation and recreation of the modern self.

Relationships are indeed essential for how the travel writer defines and imagines the self, and they are strongly related to concepts of performativity, gender

15. Doreen Massey, "A Global Sense of Place," *Marxism Today* 38 (June 1991): 24–9.

and body, as is demonstrated by Broomans and Den Toonder. The bodily act of speaking is considered a performative act that, in gender theory, establishes gender identity through "*a stylized repetition of acts*."[16] As Broomans observes, in Wollstonecraft's *Letters*, performative acts refer to the various roles that the author adopts as cultural transmitter, each role requiring a specific kind of speech and voice. Gender performativity as a ritualised socially constructed norm is evident in *Désorientale*, where the narrator struggles to accept her own body as a result of her lesbian sexual identity, which is considered a crime in the Islamic Republic. The debate on gender and sexuality stretches back to the Enlightenment's critique of Catholic teachings on the topic, as Van Tilburg demonstrates in her analysis of the attitudes towards sexuality that Bougainville reports following his contact with Pacific culture.

In addition to the concepts of travel and cultural transfer themselves, the travelling of ideas, networks and relationships, contact zones, space-time constructs and gender, sexuality and performativity can all be characterised as concepts that are essential in understanding the process of cultural transmission.

Concluding remarks

Much has been written on travel literature and on cultural transfer. In this volume, we combine the two – the genre and the act of transferring texts, knowledge of artefacts and ideas, as well as the study of cultural transfer. We focus on the narrator, who takes on the role of traveller, as well as that of the writer reporting on the journey and addressing the experience of finding oneself in a new environment. For that matter this volume demonstrates that the personal is important. The contributions to this volume show how the travel writer takes on various roles, which allows several key terms to be identified. We learned of the traveller-explorer (Bougainville, Stevenson) and the travelling-researcher (Nordenskjöld), as unwitting transmitters of ideas of the self and the other. We also identified the traveller-investigator with a cause (Wollstonecraft), presenting reflections on society and gender. Linking the act of defining the self to persona in the travelling process, we also propose to include the act of performativity in the study of travel literature and cultural transfer.

Another theme that needs to be explored in more depth is the implicit transfer of literature, ideas and even national characters (Reinhardt, Moissi), as demonstrated by the example of the theatre performances by guest directors and actors

16. Judith Butler, "Performative Acts and Gender Constitution: An Essay in Phenomenology and Feminist Theory," *Theatre Journal* 40, no. 4 (1988): 519. Author's italics.

in another country. The final key term is nomadism, which is reflected in the return from the global to the local (Martinsson) and those acts of reflexive cultural transfer in which the protagonist holds up a mirror to Western society (Wilhelm, Djavadi).

The combination of travel writing and cultural transfer thus generates new topics and approaches in research, such as the various roles the travel writer fulfils, deliberately or not, especially that of a cultural transmitter. The dynamics of the self, persona and performativity also offers an intriguing perspective. Adding to this the idea of implicit transfer and the consideration of images that accompany texts, an increasing sensitivity to multimodality in the analysis of travel writing is required. Combining various theoretical approaches in the study of travel writing and cultural transfer can offer fruitful insights into the act of mirroring. Examples include the use of historicism when analysing travel literature written in the past, or postcolonialism when examining travel literature in the twentieth and twenty-first centuries. Compelling mirroring occurs when, by means of reflexive cultural transfer, the East confronts the West with its own Occidental gaze.

Further development of approaches to travel writing will be necessary in the digital era. After the Second World War, modes of travel changed, as did the way in which travel was reported. In the twenty-first century, new social media has given rise to a more immediate mode of providing information on travel experiences. Travel has almost become akin to a fast-food experience, instantly accessed, consumed and assessed in blogs or on internet forums. While a new travel writing genre seems to have emerged, climate change, the experience of the COVID-19 pandemic and a general need to do something about the speed of life today might result in a return to a more contemplative and slow mode of travelling. The travel writer might resort to paper notebooks to write about the last glaciers or exotic flowers that can adjust to the changing environment. The future of travel and writing about it has become uncertain. It might even be an interstellar undertaking that heads towards new horizons, as is hoped for by the travellers in *Aniara*. This, however, is another story; perhaps the subject of a subsequent volume.

References

Adams, Percy G. *Travel Literature and the Evolution of the Novel*. Lexington, 1983.

doi Broomans, Petra. "Preface." In *From Darwin to Weil. Women as Transmitters of Ideas*, edited by Petra Broomans, vii–xiii. Groningen: Barkhuis, 2009.

doi Butler, Judith. "Performative Acts and Gender Constitution: An Essay in Phenomenology and Feminist Theory." *Theatre Journal* 40, no. 4 (Dec. 1988): 519–31.

Das, Nandini and Youngs, Tim (eds.). *The Cambridge History of Travel Writing*. Cambridge: Cambridge University Press, 2019, 93–107.

Forsdick, Charles, Kinsley, Zoë and Walchester, Kathryn (eds.) *Keywords for Travel Writing Studies: A Critical Glossary*. London/New York: Anthem Press, 2019.

Figes, Orlando. *The Europeans: Three Lives and the Making of a Cosmopolitan Culture*. New York: Metropolitan Books, 2019.

Holmes, Richard. *Footsteps. Adventures of a Romantic Biographer*. Harmondsworth: Penguin Books, 1985.

Lidström, Carina. *Berättare på resa. Svenska resenärers reseberättelser 1667–1829*. Stockholm: Carlson, 2015.

Massey, Doreen. "A Global Sense of Place." *Marxism Today* 38 (June 1991): 24–9.

Pratt, Mary Louise. *Imperial Eyes: Travel Writing and Transculturation*. London: Routledge, 1992.

Sandrock, Kirsten. "Discourses of Travel Writing," In *Handbook of British Travel Writing*, edited by Barbara Schaff, 31–54. Berlin/Boston: De Gruyter, 2020.

Schaff, Barbara. *Handbook of British Travel Writing*. Berlin/Boston: De Gruyter, 2020.

CHAPTER 1

Cultural transfer in the French Enlightenment

Sexuality and gender in Bougainville's and Diderot's writings on Tahiti

Marja van Tilburg
University of Groningen

Over the last decade, scholars of the Enlightenment have shown how cultural critics all over Europe – fashioning themselves as *philosophes* – took inspiration from non-Western cultures. Articulating their criticism of contemporary European society, they drew comparisons with seemingly better functioning societies in other parts of the world. How they selected and processed relevant information, however, has not been given due attention. How this information was subsequently used in Enlightenment thinking has hardly been analysed – as if *philosophes* only adopted ideas instead of examining and revising them in the process. To point out this last aspect of Enlightenment culture, this article discusses two important late eighteenth century texts. The first is the report of the French explorer Bougainville about his sojourn on Tahiti (1771). This author renders extraordinary sexual mores intelligible by referring to the Enlightenment concept of "natural man." The second is the commentary of the French *philosophe* Diderot, who used the above report to develop an alternative to French sexual mores (several versions, 1771–1784). Together, these two texts offer examples of Enlightenment cultural transfer: the explorer describing Tahitian culture in a way which inspires the cultural critic to formulate new sexual mores and to discuss the feasibility of their implementation in contemporary Europe.

Keywords: Cultural Transfer, Enlightenment, sexuality, gender, Tahiti

Cross-cultural encounters were a contributing factor to the European Enlightenment. Cultural critics – going by the name of *philosophes* – proved avid readers of travelogues and other accounts reporting on non-Western peoples, and they used the information to better formulate their criticism of contemporary Euro-

https://doi.org/10.1075/fillm.20.01van

pean society.[1] The *philosophe* Voltaire (1694–1778), for example, wrote a comparative history of the world's civilisations – starting with China and India to avoid the traditional Eurocentric approach. The comparison aimed to provide further insight into the process of establishing a functioning social order and achieving cultural advancement. In addition, it aimed to enhance critical debate in contemporary French society. His *Essay sur les mœurs et l'esprit des nations* (1756) stands in a long tradition. In the sixteenth century, the essayist Montaigne (1533–1592) had used reports on the inhabitants of the Americas to criticise religious strife in war-torn France. He had argued that the cannibalistic Tupinamba were more civilised than the warring French, claiming that while the former only abused people after they had passed away, the latter brutalised human beings while they were still alive. His essay "Des cannibals" (1580) was influential on various levels, with the argument contributing to the development of a civic – as opposed to a religious – perception of the state. This new perspective proved of crucial importance to the ending of the wars of religion in France (1562–1598).

Although a willingness to learn from non-Western peoples can be discerned well before the Enlightenment, *philosophes* nevertheless took their interest in non-Western peoples a step further. They developed novel ideas regarding humans – as recently demonstrated by the political scientist Siep Stuurman and cultural historian Anthony Pagden. Stuurman points to the development of the concept of "natural man," which denotes that which humans have in common, despite differences in physique and culture. He credits the French philosopher Poulain de la Barre (1647–1723) with developing this concept.[2] Pagden, in turn, engages with the impact of this idea on Enlightenment thinking. For him, it provides the foundation for the Enlightenment notion of equality, with Enlightenment authors supposing all humans to be equal, despite appearances to the contrary.[3] The connected ideas of humanity and equality thus fit into the general approach of Enlightenment thinkers, who looked for natural laws governing human behaviour in the same way as scientists looked for natural laws governing the cosmos. Following this approach, comparisons between non-Western peoples and Europeans could help define "human nature." Following Pagden's train of thought, it becomes apparent why non-Western societies could provide models of behaviour for Europeans.

---

1. David Greaber and David Wengrow, *The Dawn of Everything: A New History of Humanity* (New York: Farrar, Straus and Giroux, 2021), 27–77.

2. Siep Stuurman, *The Invention of Humanity: Equality and Cultural Difference in World History* (Cambridge MA: Harvard University Press, 2017), 258–63.

3. Anthony Pagden, *The Enlightenment: And Why It Still Matters* (Oxford: Oxford University Press, 2013), 35–52.

Although the above research offers significant insights into the importance of cross-cultural encounters to Enlightenment thinking, it stops short of analysing this process: The selecting and processing of information on "other" cultures has never been given due attention. Moreover, the use of this information in Enlightenment debates has hardly been analysed. To acquire insight into this complex process, this article takes inspiration from the concept of "cultural transfer" as proposed by Broomans and Klok.[4] It traces cultural transfer in eighteenth century travel writing following directions pointed out by Broomans and Den Toonder.[5] It engages with differences between actual and fictional travel writing. It takes account of specific roles of the author – actual explorer versus fictitious reporter. Above all, it focuses on the reporting of actual events and the subsequent transforming of these events into "alternative practices". This focus fits Enlightenment attention to "best practices" – instances in which non-Western societies functioned better than European ones. Usually, *philosophes* drew comparisons between non-Western and European customs to point to the dysfunctional aspects of European culture. To a lesser extent, they presented alternatives to the European tradition. It is not that they were not open to the ideas of other Europeans – Voltaire's *Essay* offers many examples of ideas pertaining to society, such as charity and tolerance. However, on the whole, they found non-European practices more useful for their purpose of critiquing and consequently improving contemporary society.

This chapter analyses a well-known example of Enlightenment engagement with "other" cultures: the late eighteenth century "discovery" of Tahiti. Before the late eighteenth century, the Pacific was little known to Europeans. For late eighteenth-century explorers, Pacific peoples behaved in strange ways, with their customs presenting an epistemological challenge.[6] Pacific women, in particular, exhibited unthinkable and unexpected behaviour.[7] Consequently, the travel accounts describing such behaviour were read avidly by a rapidly growing readership. The analysis here will be based on one of these accounts: the report of the first

4. Petra Broomans and Janke Klok, "Thinking about travelling ideas on the waves of cultural transfer," in *Travelling Ideas in the Long Nineteenth Century*, eds. Petra Broomans and Janke Klok (Groningen: Barkhuis, 2019), 9–28.

5. Petra Broomans and Jeanette den Toonder, "Introduction," in this volume.

6. Allen Frost, "The Pacific Ocean: The Eighteenth Century 'New World,'" in *Facing Each Other*, ed. Anthony Pagden, 2 Vol. (Aldershot: Ashgate, 2000), II, 591–634; Harry Liebersohn, *The Travelers' World: Europe to the Pacific* (Cambridge MA and London: Harvard University Press, 2006).

7. Serge Tcherkézoff, *'First Contacts' in Polynesia: The Samoan Case (1722–1848): Western Misunderstandings about Sexuality and Divinity* (Canberra and Christchurch: Jl. of Pacific History and Macmillan Brown Centre for Pacific Studies, 2004).

French explorer to visit Tahiti, Admiral De Bougainville (1729–1811). His travelogue strongly suggested that Tahitians followed their natural inclinations and, in particular, pursued sexual encounters without inhibitions. Consequently, Enlightenment readers came to associate Tahitians with Rousseau's idea of natural man or the noble savage.[8] The Enlightenment interest in Tahiti also inspired the French *philosophe* Diderot (1713–1784). This armchair traveller used Bougainville's travelogue to sketch the outlines of a "natural" sexual order. *En passant*, he discussed the feasibility of implementing new sexual mores in Europe. Thus, these texts offer insight into the process of cultural transfer by way of travel writing.

Before we start analysing these texts, some stylistic features should be pointed out. In Early Modern culture, the travelogue frequently included a chronological account of events, followed by a systematic overview of local society and sometimes a list of useful words in an appendix. Over time, the genre developed further varieties. Compilations of travel reports for a specific region presented one such variety. The Dutch physician Olfert Dapper (1635–1689), for example, compiled several such compendia on Africa and Asia in Dutch, German and French. Imaginary travel accounts presented another variety. European authors enjoyed envisaging imaginary societies on remote islands. Over time, More's *Utopia* (1516) and Bacon's *New Atlantis* (1626) inspired many other authors and artists. In addition, some *philosophes* used the genre to debate questions of society. Swift's *Gulliver's Travels* (1726) presented a refutation of Defoe's naïve rendering of a European beachcomber in the Pacific in *Robinson Crusoe* (1719).

In this context, Bougainville's choice to stray from the format makes sense. Rather than writing a chronological account of events, he focused on his sojourn on Tahiti. Not only does his *Voyage autour du Monde* have three extensive chapters on the island, but his account also prioritises local customs,[9] drawing analogies with ancient European religious rituals to make local practices intelligible. These features lend the travel report a rather ethnographic character. At first glance, Diderot's *Supplément au Voyage de Bougainville* seems to be a mere postscript to the above travelogue.[10] However, the work contains a long treatise consisting of five

8. Often Rousseau's phrase "l'homme naturel" is translated as "the noble savage". This last term is therefore more known as "natural man".

9. Louis-Antoine de Bougainville, *Voyage autour du Monde par la Frégate du Roi* La Boudeuse *et la Flûte* l'Étoile, ed. Jacques Proust (Paris: Gallimard; 1982); citations are taken from: Louis de Bougainville, *A Voyage round the World*, trans. John Reinhold Forster (Amsterdam and New York: N. Israel and Da Capo Press, 1967).

10. Denis Diderot, *Supplément au Voyage de Bougainville,* ed. Herbert Dieckmann (Genève and Lille: Librarie Droz and Librarie Giard, 1955); citations are taken from: Denis Diderot, *Political Writings*, trans. and ed. John Hope Mason and Robert Wokler (Cambridge: Cambridge University Press, 1992), 35–75.

distinct parts which showcase typical Enlightenment genres. Both the first and last parts contain fictional dialogues between two fictitious Frenchmen concerning the usefulness of Bougainville's report. This stylistic means is often applied in Enlightenment treatises to highlight a topic from various angles.[11] The other parts contain an imaginary travel report and also a response from a fictitious local elder to the arrival of the Europeans. While posing as travel literature, these two texts present a different culture in a confronting way. In describing an alluring world, they offer food for thought for Enlightenment audiences.[12]

Furthermore, both authors write extensively on indigenous women and sexual practices because these are important topics in Enlightenment thinking. On the whole, Enlightenment audiences furthered women's emancipation from the traditional roles of wife and mother. In addition, French Enlightenment *philosophes* criticised Christian teachings on sexuality for alienating people from human nature. In their view, this alienation was the root of various social problems such as "illicit" liaisons, "illegitimate" children and prostitution. Apparently, the authors intended to contribute to the Enlightenment debate on gender and sexuality. Throughout their texts, they refer to ideas which are central to Enlightenment discourse.[13] Last but not least, it becomes clear that Bougainville and Diderot held different positions on the Enlightenment. The explorer seeks to be taken seriously by Enlightenment intellectuals, as shown in the effort to describe Tahitian culture with reference to Enlightenment theory, as will be discussed. The *philosophe* is highly critical of Enlightenment interest in Tahiti, as will also be demonstrated. It may be inferred from their writings that Bougainville believes in the Enlightenment mission of gaining knowledge and furthering progress, whereas Diderot expresses doubts about the feasibility of this ideal and reveals an ambivalence towards his fellow travellers of the Enlightenment.

A textual analysis of Bougainville's travelogue and Diderot's commentary will pull these threads together. The analysis will prioritise how both authors describe Tahitian practices. It will show the voyager trying to convey Tahitian customs as exemplifying a "natural" way of life. It will point out how the *philosophe* uses the voyager's information to critique European sexual mores and speculate about a new sexual morality. In addition, the analysis will show how both authors use

---

11. Dena Goodman, *The Republic of Letters: A Cultural History of the French Enlightenment* (Ithaca: Cornell University Press, 1994), 187–97.

12. The presentation of information in an entertaining way is also a feature of Enlightenment literature; Robert Darnton, *The Forgotten Bestsellers of Pre-revolutionary France* (New York: Norton, 1995), 85–114.

13. Urs Bitterli, *Cultures in Conflict*, trans. Ritchie Robertson (Stanford: Stanford University Press, 1986), 155–77.

an array of ideas and images to do so. Their extraordinary efforts testify to the Enlightenment endeavour to explore the world, to learn from non-Western peoples and to pursue equality and liberty for all.

Bougainville's travelogue

Bougainville made his voyage in a distinctly imperialist context. The foreign minister, Etienne Francois de Choiseul, was seeking to compensate for the loss of the French colony in North America to the British in 1763. Because opportunities for colonisation in Asia had also dwindled, the hope was to discover new territories to colonise. A decade earlier, the geographer Charles De Brosses had rekindled interest in Ptolemy's assumption of a landmass in the southern hemisphere. His publications led Choiseul to consider an expedition to chart this "Terra Australis Incognita."[14] When Bougainville learned of this, he volunteered to organise and finance the enterprise.[15] Realising the potential for the voyage, he added a botanist to the crew to also document discoveries from a naturalist point of view. Although his engagement with scientific inquiry was genuine, the main aim of the voyage was political. The French knew that the British were already exploring the South Seas and if they wanted to establish new colonies, they would have to step up their efforts.[16] Thus, the voyage was part of the competition for wealth and empire between these two imperial powers.

Bougainville was well aware of the political aspects of the voyage. As an army officer, he had fought against the British in North America and, as a diplomat, he had negotiated the peace treaty. More than anybody, he realised the loss of New France was no small defeat. However, his motivation was also grounded in a profound interest in science. While reading law at the University of Paris he also studied the classics, linguistics and mathematics. Early in his career, he published a treatise on integral calculus, which earned him membership of the British Royal Society. In addition, he was familiar with the Enlightenment – the cultural movement furthering scientific inquiry and using knowledge to improve society. His parents were acquainted with the *philosophe* d'Alembert – co-editor of the Enlightenment's most seminal publication, the *Encyclopédie* (1751–1777).[17] Many Enlightenment themes will have been discussed over dinner. Without doubt, the explorer shared in the intellectual interests of his day as much as in the political domain.

14. Bitterli, *Cultures*, 158–9.

15. Michael Ross, *Bougainville* (London and New York: Gordon & Cremonesi, 1978), 85.

16. Bitterli, *Cultures*, 161–2.

17. Ross, *Bougainville*, 10, 15.

The explorer published his travelogue relatively soon after his return to France, in 1771. As already indicated, in the chronological account of the voyage the chapters on Tahiti stand out, as the text presents local customs rather than daily events. The descriptions are detailed and precise, as we will see. In addition, they show Bougainville's intellectual interests, with the author creating comparisons between Tahiti and Ancient Rome. While drawing these comparisons, he systematically refers to Enlightenment notions of human nature. Reading between the lines, there is also an apparent awareness of women's subservient role in society. In addition, the style is lively and engaging, as if to underline the extraordinariness of the experience. This can be said especially for the passages reporting on sexual encounters with local women. Of course, these drew the attention of his contemporary readers and fuelled public interest in the exploration of the Pacific.

For a long time, historians considered *Voyage autour du Monde* to be an example of Enlightenment ideology, with the explorer thought to have depicted Tahitians according to Rousseau's image of noble savages. More recently, the historian Anne Salmond has argued for the influence of more traditional European imagery. She notes how much Bougainville's representation of Tahiti resembles ancient Arcadia and compares the description of the island to contemporary paintings of this utopia of Greek mythology.[18] The literary critic Vanessa Smith has argued that the idyllic aspect is the result of Bougainville's style, with the explorer suggesting ready connections by offering comprehensible translations. For example, upon arrival, the French ships were met with a flotilla of canoes filled with locals, all shouting "tayo."[19] Bougainville translates this word as "friend,"[20] which suggests that the cross-cultural encounter was easy to comprehend.[21] Thus, the explorer also suggests that it was easy to make such a connection with the Tahitians.

Despite the effort to render Tahitians as comprehensible and relatable, however, Bougainville cannot conceal finding some aspects of Tahitian culture problematic. Most pertain to women's sexual behaviour towards the French. He frames this behaviour as part of some indigenous ritual – probably to render the event comprehensible to the readership.[22] Describing the actual practice, he points to differences with European sexual practices – sexual encounters taking place in

18. Anne Salmond, *Aphrodite's Island: The European Discovery of Tahiti* (Berkeley: University of California Press, 2010), 20.

19. Vanessa Smith, *Intimate Strangers: Friendship, Exchange and Pacific Encounter* (Cambridge: Cambridge University Press, 2010), 1–2.

20. Bougainville, *A Voyage*, 217; in original: "ami"; *Voyage*, 225.

21. Smith speaks of the "universal translatability" of the word "tayo"; Smith, *Intimate Strangers*, 2.

22. The narrative strategy to present a given ethnic practice using a familiar Western concept is also used by contemporary anthropologists, as pointed out by Clifford Geertz, *Works and Lives: The Anthropologist as Author* (Stanford: Stanford University Press, 1988), 49–72.

public places, in large gatherings, accompanied by music, etc. The first example is taken from the above-mentioned description of the arrival at Tahiti. The French ships were met by canoes filled with people shouting words of welcome. Among them were young women in a state of undress because their sarongs had been pulled off by fellow passengers. According to the account, the young women were presented to the accidental visitors as a gift. The passage finishes with the story of one crew member who jumped from the ship and invited one of the girls to meet him ashore. The moment he landed on the beach, he was met by local men and was undressed and inspected thoroughly. Only then, was he allowed to continue and "... content those desires which had brought him on shore with her."[23] However, the reception had scared him so much that he immediately returned to the safety of the ship.[24] This incident may have been added to the report in order to convince the readership that the crew was not mistaken. The young women were put on show for them to admire and desire. This anecdote was meant to convince the readers of the Tahitian's enthusiasm in welcoming the French.

While the above passage presents an incident where the French were "just looking," the next example showcases an actual sexual encounter between French men and Tahitian women:

> Our people were daily walking in the isle without arms, either quite alone, or in little companies. They were invited to enter the houses, where the people gave them to eat; nor did the civility of their landlords stop at a slight collation, they offered them young girls; the hut was immediately filled with a curious croud of men and women, who made a circle round the guest, and the young victim of hospitality. The ground was spread with leaves and flowers, and their musicians sung an hymeneal song to the tune of their flutes. Here Venus is the goddess of hospitality, her worship does not admit of any mysteries, and every tribute paid to her is a feast for the whole nation.[25]

23. Bougainville, *A Voyage*, 219; in original: "... contenter les désirs qui l'avaient amené à terre avec elle," *Voyage*, 227.

24. Bougainville, *A Voyage*, 219; in original: *Voyage*, 227.

25. Bougainville, *A Voyage*, 227–8; "Chaque jour nos gens se promenaient dans le pays sans armes, seuls ou par petites bandes. On les invitait à entrer dans les maisons, on leur y donnait à manger; mais ce n'est pas à une collation légère que se borne ici la civilité des maîtres de maisons; ils leur offraient des jeunes filles; la case se remplissait à l'instant d'une foule curieuse d'hommes et de femmes qui faisaient un cercle autour de l'hôte et de la jeune victime du devoir hospitalier; la terre se jonchait de feuillage et de fleurs, et des musiciens chantaient aux accords de la flûte une hymne de jouissance. Vénus est ici la déesse de l'hospitalité, son culte n'y admet point de mystères, et chaque jouissance est une fête pour la nation," *Voyage*, 235. Later in the report, Bougainville uses the word "l'hyménée," *Voyage*, 236, when describing a similar event. This is probably a reference to the hymen.

Festive gathering at Tahitian home 1773. Etching/engraving
Print made by: Francesco Bartolozz. After: Giovanni Battista Cipriani.
More information on BM-site

This passage presents sexual intercourse taking place as part of a customary reception with ritualistic aspects. Many locals gather at the house and participate in the event. Every local is familiar with the proceedings. Furthermore, the representation points to the differences from European ways, with the people positioning themselves in a circle around the couple. The intercourse takes place on a bed of flowers and is accompanied by music. Even the reference to the goddess Venus – probably the only aspect of the ritual familiar to the European readership – stresses difference. In Tahiti, she is the goddess of hospitality. However, while the reference is made, of course, to veil the actual act of sexual intercourse, by making this reference, the explorer frames the intercourse as part of a rite. Somewhat later in the section, he comments on the young woman presented to the French, as we will see. For now, it is important to note the communicative aspect. While the Tahitians are presented as "other," their strange ways are communicated precisely as well as in a lively manner.

As mentioned, the passages describing these sexually charged events are also passages describing women. In contrast to the exotic rendering of indigenous customs, the women are depicted in a relatively familiar fashion. This characteristic is apparent in the description of the welcoming ceremony mentioned above:

> The periaguas were full of females; who, for agreeable features, are not inferior to most European women; and who in point of beauty of the body might, with much reason, vie with them all. Most of these fair females were naked; for the men and the old women that accompanied them, had stripped them of the garments which they generally dress themselves in. The glances which they gave us from their periaguas, seemed to discover some degree of uneasiness, notwithstanding the innocent manner in which they were given; perhaps, because nature has every where embellished their sex with a natural timidity; or because even in those countries, where the ease of the golden age is still in use; women seem least to desire what they most wish for.[26]

The first sentence presents Tahitian women as being as beautiful as European women. In the next sentence in the French original, the women are described as "nymphes" – ancient goddesses who are customarily depicted nude. This phrasing indicates to the well-educated reader that the women were naked, before the author reveals this fact in so many words. Considering the strong humanist tradition in European culture, the simile creates an association with Antiquity as well. After describing the women's appearance, the author turns to their behaviour. First, he mentions their naivety and timidity – thus presenting the young women as "normal" to the readership. Subsequently, he offers two interpretations of the women's behaviour.

The first explanation draws on the Enlightenment theory of natural laws, which determine human behaviour in a similar way to the law of gravity determining the cosmos. Following this approach, it is some perceived natural law that makes young women shy away from sexuality. The second explanation refers to a historical approach developed within Enlightenment circles. This proposes that societies become more complex and more civilised over time. The implied assumption of progress had been critiqued by Rousseau, who argued that socioeconomic development generated estrangement from nature. Following Rousseau's line of reasoning, Tahiti had remained a simple society and was, by consequence, still close to nature. Here, it should be noted that these two approaches dominated contemporary thinking on society and civilisation.[27] In

---

26. Bougainville, *A Voyage*, 217–8; "Les pirogues étaient remplies de femmes qui ne le cèdent pas pour l'agrément de la figure au plus grand nombre des Européennes, et qui, pour la beauté du corps, pourraient le disputer à toutes avec avantage. La plupart de ces nymphes étaient nues, car les hommes et les vieilles, qui les accompagnaient, leur avaient ôté la pagne dont ordinairement elles s'enveloppent. Elles nous firent d'abord, de leurs pirogues, des agaceries où, malgré leur naïveté, on découvrait quelque embarras; soit que la nature ait partout embelli le sexe d'une timidité ingénue, soit que, même dans les pays où règne encore la franchise de l'âge d'or, les femmes paraissent ne pas vouloir ce qu'elles désirent le plus," *Voyage*, 225–6.

27. Stuurman, *Invention*, 258–344.

the context of the travelogue, these explanations rendered the women's behaviour comprehensible. All in all, the author goes to great lengths to present Tahitian women as fairly civilised and their behaviour as somehow understandable to Europeans. This representation appealed to the contemporary readership and also offered Diderot food for thought, as we will see.

The same tension between an exotic ritual and comprehensible femininity comes to the fore in the description of Tahitian hospitality mentioned above. While describing the proceedings of the ritual, Bougainville refers to the customary sexual encounter as a duty and the young woman involved as a victim of tradition. He acknowledges that tradition prescribes certain practices which leave the young woman with no choice. Very briefly – in half a sentence – he also presents the young woman as a person who might have preferred a different course of events. This phrasing may have rendered the proceedings more relatable to many readers. Enlightenment circles had encouraged women to emancipate themselves from the subservient role that they had in European society. For example, women were allowed to participate in their meetings. A new etiquette had been introduced to ensure that men treated women respectfully. Furthermore, the Enlightenment identified strongly with its treatment of women. In the words of Voltaire, the above-mentioned "commerce between the sexes" epitomised civilisation.[28] By emphasising that the young woman was obliged by tradition, Bougainville appeals to this tendency within Enlightenment culture. *En passant*, he communicates that he shares the Enlightenment attitude towards women.

The above reading of the descriptions of Tahitian women participating in indigenous sexually charged practices reveals Bougainville to be of two minds. Describing local customs, the explorer tends to stress difference. In the first scene, he applies the stylistic device of exaggeration: the flotilla is very large, the people are very enthusiastic, and their behaviour is very inviting. In the second scene, he points out differences to European practices: the ritualistic aspects, the joyous audience and the festive character. Describing women's roles, he attempts to present their behaviour as intelligible. He creates analogies to ancient goddesses – situating the women in a bygone but familiar civilisation. In the description of the welcoming ceremony, he adds references to Enlightenment thinking regarding sexuality. Basically, he renders women as "civilised" and their behaviour as "natural." In addition, in the description of the reception in Tahitian homes, he points to the tension between cultural custom on the one hand and individual preference on the other. This echoes Enlightenment concerns with women's place in society. Overall, he shows himself to share Enlightenment views on sexuality and femininity.

28. Cited by Goodman, Republic, 4, 9–10.

Young Tahitian woman in ceremonial attire 1785. Engraving
Print made by: John Keyse Sherwin. After: John Webber.
More information on BM-website

From the perspective of cultural transfer, Bougainville's report on Tahitian sexual practices exhibits some ambiguity. His description of some specific practices offers sufficient information for European readers to understand the proceedings during the French sojourn. On the whole, he presents these practices as different from European mores. However, the descriptions of the women present exceptions to this rule. The writing uses imagery taken from ancient mythology and well-known ideas from Enlightenment discourse. Although Europeans were familiar with these images and ideas, they remained abstractions, which functioned as a screen behind which the women remained hidden. From this, it can be inferred that Bougainville had difficulty perceiving the Tahitian women as specimens of "natural man." That the voyager tried to depict Tahitian women as such – as "normal" from an Enlightenment perspective – shows that he sought to be a man of the Enlightenment.

Diderot's commentary

When Bougainville published his travelogue, Diderot was already a central figure in European Enlightenment circles. His position in this cultural movement was mainly based on his contribution to the *Encyclopédie, ou dictionnaire raisonné des sciences, des arts et des métiers.* He had developed the format of the new genre with the above-mentioned D'Alembert. Many scientists and *philosophes* contributed to this serial publication intended to disseminate knowledge to a large audience.[29] As editor, he may have been the most informed and best connected of all.

Writing his commentary, the *philosophe* engaged with the travelogue on two levels, addressing Tahitian culture and what the Enlightenment could learn from it. He also addressed the reception of the report in Enlightenment circles and conveyed his criticism in the content as well the phrasing. The outcome is a very layered text that dazzled the readership with intellectual insight and wit. Diderot's engagement with the themes of exploration and colonisation is a constant thread in his writing. In the context of the *Encyclopédie*, he had written on the impact of colonisation on the native peoples of America and India. Ultimately, this resulted in an extensive comparative study written in collaboration with fellow *philosophe* Raynal, called *Histoire philosophique et politique des deux Indes* (1772–1781). Writing his commentary, he also engaged with other Enlightenment concerns, playing with the concept of "natural man" by developing ideas on "natural woman," for example. In this context, he speculated about natural sexuality as opposed to Catholic teachings on this topic. Here, the influence of his Jesuit education came to the fore. From the start, Diderot feared that the commentary might not be well received by the public – neither the traditional elite nor the Enlightenment readership. He revised the text at least five times, and the manuscript was only published posthumously.

The *Supplément au voyage de Bougainville* has five parts. Part 2 includes a monologue by an elderly Tahitian man lamenting the loss of autonomy that his people will suffer following their "discovery." Parts 3 and 4 present a drawn out dialogue between a Tahitian chief and a French chaplain regarding Tahiti's sexual mores. This dialogue has an *entre-act*, in which two Frenchmen discuss dysfunctional aspects of Western sexual mores – using a widely publicised court case as an example. The first and last parts present dialogues between the same Frenchmen, addressing questions which the average reader of the travelogue might have – such as the reliability of Bougainville's account. Most of the dialogue addresses the applicability of Tahitian sexual mores to contemporary France. All five parts

29. Daniel Brewer, *The Discourse of Enlightenment in Eighteenth-Century France: Diderot and the Art of Philosophizing* (Cambridge: Cambridge University Press, 1993), 13–55.

make reference to Bougainville's report. The chief's explanation of Tahitian sexual mores offers details which are clearly copied. At the same time, the chief's narrative is consistent with Enlightenment discourse on "natural man," as demonstrated by Stuurman. Clearly, the *philosophe* presents a rendering of the explorer's account from the perspective of Enlightenment thinking. Here, Diderot evidences what Pagden claimed in particular: that the *philosophes* used non-Western cultures as food for thought.

The dialogue between the chief and the chaplain presents an implicit comparison between Tahitian and European sexual mores. In the opening sentence, the chief offers his youngest daughter as a companion for the night. The chaplain declines the invitation, explaining that as a monk he has to adhere to the vow of chastity. After the monk has outlined French sexual mores, the chief tells him that islanders can choose their partners freely. Marital unions last for only a month so that men and women can change partners, while society as a whole takes care of the offspring. As indicated above, some elements in this dialogue are taken from Bougainville's report, such as the offer of a female companion for the night. Most aspects are made up by the author in order to create an implicit contrast between two sexual regimes. A closer look shows the contrast to be created from the perspective of naturalist theory. It presents a sexual order which is in accordance with natural law, one which ensures procreation and allows at the same time for the fickleness of the human heart. More importantly, Tahitian society acts responsibly towards its progeny while allowing women and men personal freedom. Clearly, the dialogue aims to make apparent to the readership that French sexual mores go against nature.

In the dialogue, Diderot not only critiques French sexual mores but also European perceptions of sexuality. In the same dialogue, the author has chief Orou explain how island women look for strong partners and how men prefer a woman "who promises many children [...] children who will be active, intelligent, courageous, healthy and robust."[30] These remarks suggest that Tahitian desire is connected to procreation. In addition, the author addresses the topic in more general terms in the last sentence of the paragraph: "The Venus of Athens and that of Tahiti have next to nothing in common."[31] Here, the reference to Venus draws our

30. Diderot, *Political Writings*, 56; "qui promet beaucoup d'enfants [...] et qui les promets actifs, intelligens, courageux, sains et robustes," *Supplément*, 35–37.

31. Diderot, *Political Writings*, 56; "Il n'y a presque rien de commun entre la Vénus d'Athenes et celle d'Otaïti," *Supplément*, 36; the editor, Herbert Dieckmann, explains that in some manuscripts the citation is followed by the following sentence: "L'une est Venus galante; l'autre est Venus féconde." (The one is gallant Venus; the other is fertile Venus); Dieckmann attributes this sentence to an eighteenth-century editor of Diderot's work named Naigeon, *Supplément*, 36. Because this addition may not be Diderot's, I have left it out of the analysis.

attention. It is most likely that Diderot used this simile because the explorer also used it.[32] In any case, he uses it in the same way – to indicate a sexual encounter. In both texts, the word "Venus" is supposed to say it all. At the same time, the author is creating difference: the Venus of Tahiti aims at generating talented children, as Orou has explained. This interpretation implies naturalist theory, with the Tahitians depicted as concerned with the quality of their progeny and the future of society. The Venus of the Europeans signifies the opposite – Europeans have only personal gratification on their minds.

As already mentioned, the exploration of the Pacific caught the attention of the European public. Diderot supposed that this was due to the reports about the local women. Several French and English explorers had published travel accounts, and most highlighted their allure and accessibility. Of course, these voyagers had interpreted the women's attitude towards sexuality as an aspect of indigenous culture.[33] However, in the above dialogue, Diderot offers a different interpretation. Chief Orou explains that his people did not intend to please the French, contrary to what the latter thought. They were after French seed, so that their women would bear children of another ancestry. The chief explains that the "French" children were intended for their feudal overlord – the levy being part of a peace settlement with a neighbouring island. In this way, the Tahitians could keep their own children. The chief acknowledges that the French were thereby to be stripped of their valuable goods, but reminds the latter that they were given pleasure in the exchange. To conclude the discussion, Diderot has the chaplain express astonishment. No doubt, the author tries to elicit the same emotion from the readership.

The above passage offers an inversion of Bougainville's description of Tahiti. Whereas the explorer presents Tahiti in splendid isolation, the *philosophe* depicts the island in the larger context of inter-island competition for wealth and power. Furthermore, whereas the first presents sexual encounters as an integral part of Tahitian ceremonies, the latter depicts them as a type of exchange. That Diderot means "business" is apparent in the phrasing:

> On your arrival, we let you have our wives and daughters. You showed surprise and such gratitude that we laughed. You thanked us for having placed on you and your companions the greatest of all impositions. We didn't ask you for your money; we didn't loot our ships; we cared nothing for your produce; but our wives and daughters drew blood from your veins.[34]

---

32. Bougainville, *A Voyage*, 228; *Voyage*, 235.

33. Roy Porter, "The exotic as erotic: Captain Cook in Tahiti," in *Exoticism in the Enlightenment*, eds. George Rousseau and Roy Porter (Manchester: Manchester U.P., 1990), 117–44.

34. Diderot, *Political Writings*, 63; Only the last sentence is discussed: "Nous ne t'avons point demandé d'argent; nous ne nous sommes point jetté sur tes merchandises; nous avons méprisé

These sentences focus on merchandise and exchange. The Tahitians had no interest in French commodities; they were only interested in French seed. In this sentence, bodily fluids are put on a par with trade goods. More importantly, the Tahitians appear economically savvy. They are aware of their interests and know how to strike a deal. In addition, they appear to be the better traders as they pursue what they want, while the French are unaware that any exchange is taking place. Thus, the author transforms Bougainville's idyll into a world of exchange and trade, war and power – between Pacific Islanders and the French as much as among Pacific Islanders. In doing so, Diderot confronts the European readers with their prejudice regarding non-Western peoples and their false sense of superiority in international relations.

Diderot's commentary offers other examples of scepticism towards European engagement with the exploration of the South Seas. One example demonstrates how the author confronts the readers with their expectations regarding Tahiti. Here, he seems to copy Bougainville's account of sexually charged ceremonies, except he changes the function from a rite of hospitality to a rite of initiation. He also presents two versions of the ritual, one for boys and one for girls:

> The [day of] emancipation for a boy or girl is a great holiday. If it's a girl, the young men assemble around her hut the evening before, filling the air the whole night long with their singing and the sound of musical instruments. On the appointed day, she's led by her father and mother into an enclosure where there's dancing and exercises of jumping, wrestling, and running. The naked man is displayed before her, from all sides and in all attitudes. If it's a boy, then the girls undertake to please and do the honours of the ceremony, presenting before his eyes the nude female body, without reserve or furtiveness. The rest of the ceremony is enacted on a bed of leaves, as you saw on your arrival here.[35]

The above passage sketches a "rite of passage" which should teach youth about sexuality. Educating young people about responsible sexual behaviour is a recur-

---

tes denrées; mais nos femmes et nos filles sont venues exprimer le sang de tes veines" (spelling as in Diderot's manuscript), *Supplément*, 46.

35. Diderot, *Political Writings*, 55; "C'est une grande fête que le jour de l'émancipation d'une fille ou d'un garçon. Si c'est une fille; la veille, les jeunes garçons se rassemblent en foule autour de la cabane et l'air retentit pendant toute la nuit du chant des voix et du son des instrumens. Le jour elle est conduite par son pere et par sa mere dans une enceinte où l'on danse et où l'on exercise du saut, de la lutte, et de la course déploie l'homme nud devant elle sous toutes les faces et dans toutes les attitudes. Si c'est un garcon, ce sont les jeunes filles qui font en sa présence les frais et les honneurs de la fête et exposent à ses regards la femme nue, sans reserve et sans secret. Le reste de la cérémonie s'acheve sur un lit de feuilles, comme tu l'as vu à ta descente parmi nous" (spelling as in Diderot's manuscript), *Supplément*, 34–5.

rent theme in Enlightenment pedagogy.[36] Basically, Diderot offers an example of sexual education among people who live and love "naturally." Furthermore, the passage offers a distinct version for each sex. First, he points to a difference with regard to supervision. Whereas the girl's initiation is watched over by parents, the boy's ceremony is not. Next, he points to a difference with regard to the roles at the ceremonies. While the boys show their athletic prowess, the girls "do the honours." Here, Diderot engages with the Enlightenment's attention to gender. At the heart, his description fits the reader's expectations regarding sexual roles. Moreover, the author creates difference in the phrasing: whereas the boy's role is described precisely, the girl's role is sketched vaguely, they merely "do the honours." For the reader to grasp sexual education in Tahiti, they must imagine "the honours" themselves. Or, more precisely, the *philosophe* makes the reader project their own ideas (read: fantasies) into the text. Basically, he titillates the reader with anticipation regarding travelogues on Tahiti.

From the perspective of cultural transfer, it is apparent that the *philosophe* has made use of Bougainville's report. At the same time, he rewrites Bougainville's descriptions in accordance with Enlightenment discourse. The Tahitians respect human nature, having adapted the social order to human desire. They accept sexual activity as necessary to sustain society, as society as a whole takes responsibility for the offspring. True order arises from this sexual regime, as the social fabric cannot be torn apart by "secret" liaisons or "illegitimate" offspring. Here, the Enlightenment aim of creating a social order which is in accordance with human nature becomes apparent. In addition, the Enlightenment assumption shows an alignment of common good and individual interest. All in all, Diderot uses Bougainville's travelogue to depict a philosophically correct version of "natural man" and "natural woman."[37]

Thus, it can be argued that the author uses Bougainville's report to take Enlightenment debates on sexuality one step further. *En passant*, he shows himself superior to his fellow *philosophes* as his commentary addresses several different questions – from the personal to the colonial. In addition, he confronts Enlightenment readers who – so he suspects – have read the reports of the South Seas above for the exotic rituals featuring erotic scenes. Thus, he presents himself as much a *philosophe* as a critic of the cultural movement which he served throughout his life.

36. Michel Foucault, *Histoire de la Sexualité,* 3 Vol. (Paris: Gallimard, 1976–1984), I, 72–98; Marja van Tilburg, *Hoe hoorde het?: Seksualiteit en partnerkeuze in de Nederlandse adviesliteratuur 1780–1890* (Amsterdam: Het Spinhuis, 1998), 97–111.

37. Stuurman, *Invention*, 258–9.

Conclusion

The above analysis of Bougainville's travelogue and Diderot's commentary shows the complexity of the process of cultural transfer. Two distinct forms of cultural transfer come to the fore in these texts. First, the explorer tries to render Tahitian culture comprehensible to European readers. He borrows from different traditions within European culture, projecting the traditional idea of the pastoral idyll and referring to ancient mythology, and he also applies the Enlightenment concept of "natural man." Careful textual analysis shows this author to be of two minds: While describing Tahitian customs, he presents factual information on actual events, but at the same time he creates implicit contrasts to European culture. Evidently, he cannot escape "othering" Tahitians. Describing Tahitian women's behaviour, he draws on Enlightenment thinking regarding human nature. En passant, he creates analogies with ancient European icons. Thus, on an abstract level, he can point to a humanity common to Tahitians and Europeans.

Second, the *philosophe* uses Bougainville's travelogue to think through his critique of contemporary society. Drawing on the voyager's representation of Tahitian customs, he creates a drawn out contrast between fictitious islanders and contemporary Europeans. This contrast centres on the concept of natural law, where the islanders are depicted as living in harmony with human nature and Europeans not. *En passant*, the author criticises contemporary attitudes towards Tahitians. He targets the perception of Tahiti as isolated and autarchic – depicting his fictitious island as enveloped in a struggle for wealth and empire, as any other state in the eighteenth century. Furthermore, he confronts the European readership with their self-centred and shallow interest in Tahiti. In this layered text, the *philosophe* uses Bougainville's representation of Tahitian customs to sketch a new, more "natural" way of life. Essentially, he borrows from Bougainville's representation of Tahitian women to explicate a new sexual morality.

In summary, the above analysis points out Bougainville's observations of Tahitian customs, and Diderot's using Bougainville's writing to speculate about a new sexual regime. It demonstrates cultural transfer taking place: the explorer refers to the Enlightenment concept of "natural man" to explain Tahitian women's sexual behaviour. The *philosoph*e uses Bougainville's representation of Tahitian women's behaviour to outline a better functioning society. The cultural transfer shows particularly on the level of practices: the former focused on Tahitian ritual, the latter on sexual relationships among "natural men and women". In this circuitous way these two authors made Tahitian society relevant to contemporary European discourse.

References

Bitterli, Urs. *Cultures in Conflict.* Translated by Ritchie Robertson. Stanford: Stanford University Press, 1986.

Bougainville, Louis-Antoine de. *Voyage autour du Monde par la Frégate du Roi* La Boudeuse *et la Flûte* l'Étoile. Edited by Jacques Proust. Paris: Gallimard, 1982. Citations in English are taken from: Louis de Bougainville. *A Voyage Round the World.* Translated by John Reinhold Forster. Amsterdam and New York: N. Israel and Da Capo Press, 1967.

Brewer, Daniel. *The Discourse of Enlightenment in Eighteenth-Century France: Diderot and the Art of Philosophizing.* Cambridge: Cambridge University Press, 1993.

Broomans, Petra and Janke Klok. "Thinking about travelling ideas on the waves of cultural transfer." In *Travelling Ideas in the Long Nineteenth Century*, edited by Petra Broomans and Janke Klok, 9–28. Groningen: Barkhuis, 2019.

Darnton, Robert. *The Forgotten Bestsellers of Pre-revolutionary France.* New York: Norton, 1995.

Diderot, Denis. *Supplément au Voyage de Bougainville.* Edited by Herbert Dieckmann. Genève and Lille: Librarie Droz and Librarie Giard, 1955. Citations in English are taken from: Denis Diderot. *Political Writings.* Translated and edited by John Hope Mason and Robert Wokler. Cambridge: Cambridge University Press, 1992, 35–75.

Foucault, Michel. *Histoire de la Sexualité*, 3 Vol. Paris: Gallimard, 1976–1984.

Frost, Allen. "The Pacific Ocean: The Eighteenth Century 'New World.'" In *Facing Each Other*, edited by Anthony Pagden, Vol. 2, 591–634. Aldershot: Ashgate, 2000.

Goodman, Dena. *The Republic of Letters: A Cultural History of the French Enlightenment.* Ithaca: Cornell University Press, 1994.

Greaber, David and David Wengrow, *The Dawn of Everything: A New History of Humanity.* New York: Farrar, Straus and Giroux, 2021.

Liebersohn, Harry. *The Travelers' World: Europe to the Pacific.* Cambridge MA: Harvard University Press, 2006.

Pagden, Anthony. *The Enlightenment: And Why It Still Matters.* Oxford: Oxford University Press, 2013.

Porter, Roy. "The exotic as erotic: Captain Cook in Tahiti." In *Exoticism in the Enlightenment*, edited by George Rousseau and Roy Porter, 117–144. Manchester: Manchester University Press, 1990, 117–144.

Ross, Michael. *Bougainville.* London and New York: Gordon & Cremonesi, 1978.

Salmond, Anne. *Aphrodite's Island: The European Discovery of Tahiti.* Berkeley: University of California Press, 2010.

Smith, Vanessa. *Intimate Strangers: Friendship, Exchange and Pacific Encounters.* Cambridge: Cambridge University Press, 2010.

Stuurman, Siep. *The Invention of Humanity: Equality and Cultural Difference in World History.* Cambridge MA: Harvard University Press, 2017.

Tcherkézoff, Serge. *'First Contacts' in Polynesia: The Samoan Case (1722–1848): Western Misunderstandings about Sexuality and Divinity.* Canberra and Christchurch: Jl. of Pacific History and Macmillan Brown Centre for Pacific Studies, 2004.

Tilburg, Marja van. *Hoe hoorde het?: Seksualiteit en partnerkeuze in de Nederlandse adviesliteratuur 1780–1890.* Amsterdam: Het Spinhuis, 1998.

CHAPTER 2

Cultural transfer as a performative act in Mary Wollstonecraft's *Letters written during a short residence in Sweden, Norway and Denmark* (1796)

Petra Broomans
University of Groningen

Mary Wollstonecraft (1759–1797) is well known for her feminist pamphlet *A Vindication of the Rights of Women* (1792). Wollstonecraft was also an experienced traveller. She travelled to Portugal, and she lived and worked in Ireland, London and Paris. Her travel account about her stay in Scandinavia, *Letters written during a short residence in Sweden, Norway and Denmark*, was published in 1796.

Her life and works have fascinated many artists, writers and scholars over time, starting with her husband, the philosopher William Godwin (1756–1836), who published the *Memoirs of the Author of 'The Rights of Woman'* in 1798. More recently, Nigel Leask (2019), Anca-Raluca Radu (2020), Elizabeth Zold (2023), Michael Meyer (2023) and Luisa Simonutti (2024), amongst others, discussed the *Letters* in different contexts. I will begin this chapter by giving a brief overview of the outlines in this remarkable renaissance of 'Mary Wollstonecraft studies'.

I will continue by positioning Wollstonecraft's *Letters* within the genre of travel writing. In my analysis I will focus on two concepts that determine Wollstonecraft as a traveller-cultural transmitter: performativity and persona. This chapter demonstrates that the combination of persona (the self) and performativity (as writer, observer and scholar) with the approach of cultural transfer expands our understanding of the *Letters*.

Keywords: Mary Wollstonecraft, travel writing, Scandinavia, cultural transfer, performativity, persona, ethnotype

https://doi.org/10.1075/fillm.20.02bro

In the twentieth century, Claire Tomalin (1974) and Janet Todd (1976) have written seminal studies about Wollstonecraft from different perspectives. In addition, Richard Holmes published an edited version of Mary Wollstonecraft's travel letters (henceforth *Letters*) in 1987, including the biography William Godwin wrote of his wife Mary Wollstonecraft in 1798. Holmes added an illuminating introduction and notes. For my analysis of the *Letters*, I used Holmes' edition that I still consider as a splendid, though often neglected, contribution to the study of Wollstonecraft's *Letters*. In the last decades, Mary Wollstonecraft continued to feature in papers at conferences, in articles and books, often in the context of women's and gender studies or travel writing. In 2017, Brenda Ayres published her study on the biographies that were written on Mary Wollstonecraft in the last two hundred years.[1] Ayres demonstrates that Wollstonecraft is an inspiring author over time. Following the publication of Ayres' book, a revival of writing about Wollstonecraft once again occurred. Around 2020, the contours of a renaissance of Mary Wollstonecraft studies in which the *Letters* are central, became visible. Within a time frame of five years several studies were published. Nigel Leask contributed with "Eighteenth-Century Travel Writing" with a substantial part on the *Letters* in *The Cambridge History of Travel Writing* (2019), Anca-Raluca Radu wrote about the *Letters* in *Handbook of British Travel Writing* (2020), Elizabeth Zold and Michael Meyer both published on the *Letters* in *Traveling Bodies. Interdisciplinary Perspectives on Traveling as an Embodied Practice* (2023), and Luisa Simonutti paid attention to the *Letters* in the volume *Gender and Cultural Mediation in the Long Eighteenth Century. Women across Borders* (2024). The topics they write about are manifold.

Like Holmes, Leask regards the *Letters* "as the triumphant culmination of a century of travel writing, representing a Romantic synthesis of detailed observation, philosophical reflection, social critique, and 'affective realism'."[2] According to Leask, Wollstonecraft uses different narratives: for her philosophical reflections and her emotional feelings she applies an episodic structure. In her investigation of the state of art of the development in Scandinavia, "she adopts from Forster a programmatic concern for disciplined inquiry balanced with affective judgement, placing the 'modern' self of the traveller in a dialectical relationship" with what Wollstonecraft considered undeveloped countries.[3] Leask comments that for

1. Brenda Ayres. *Betwixt and Between: The Biographies of Mary Wollstonecraft.* London, New York and Delhi: Anthem Press, 2017.

2. Nigel Leask. "Eighteenth-Century Travel Writing," in *The Cambridge History of Travel Writing*, edited by Das, Nandini and Youngs, Tim. Cambridge: Cambridge University Press; 2019, 105.

3. Leask, "Eighteenth-Century Travel Writing", 105.

Romantic writers such as Mary Wollstonecraft, "science and 'pure imagination' were the Scylla and Charybdis through which the integrated travel accounts" had to be told.[4]

The article by Anca-Raluca Radu in *Handbook of British Travel Writing* (2020) on Mary Wollstonecraft's travel account, describes the biographical context of the *Letters.*[5] Radu gives a short state of the art regarding previous work on Wollstonecraft and describes the route Wollstonecraft took in Scandinavia. One of Radu's theses is that "Wollstonecraft performs several scripted roles and identities in the *Letters*"[6] and that, in doing so, Wollstonecraft adopts ambiguous roles. In the *Letters*, she argues, Wollstonecraft is a victim as well as a heroine "she is a proud and exacting woman, weak and fragile, in agony of being abandoned and relentlessly demanding confirmation of familial prospects from Imlay."[7] Wollstonecraft declares that she will use the first person and, as Radu states, on the one hand she is aware of her "self-centredness," and, on the other, is confident of her writing strategy and "qualities of her own writing."[8]

In her contribution "Motherhood and the Embodied Traveler in Wollstonecraft's Letters Written during a Short Residence in Sweden, Norway, and Denmark", Elizabeth Zold focuses on Wollstonecraft as a single mother travelling with her one year old daughter Fanny and Marguerite, the nanny. The central theme is the construction of motherhood and the body. To be a mother is one of Wollstonecraft's roles: "Within her travelogue, Wollstonecraft writes from a multitude of subjectivities: mother, slighted lover, philosopher, traveler, and feminist, and the text reflects the concerns of each of them."[9] The different roles are reflected in the narratives Wollstonecraft applies in the *Letters.* Zold regards Wollstonecraft as an emotional traveller, who connects the emotional expression with reason. Zold emphasises the relation between Wollstonecraft and her daughter as well as the embodiment.

4. Leask, "Eighteenth-Century Travel Writing", 100.

5. Anca-Raluca Radu, "Mary Wollstonecraft, Letters Written in Sweden, Denmark and Norway, 1796," in *Handbook of British Travel Writing*, ed. Barbara Schaff (Berlin: de Gruyter, 2020), 267–95.

6. Radu, "Mary Wollstonecraft, " 271.

7. Radu, "Mary Wollstonecraft," 271.

8. Radu, "Mary Wollstonecraft," 271.

9. Elisabeth Zold, "Motherhood and the Embodied Traveler in Wollstonecraft's Letters Written during a Short Residence in Sweden, Norway, and Denmark," in *Traveling Bodies. Interdisciplinary Perspectives on Traveling as an Embodied Practice*, eds. Nicole Maruo-Schröder, Sarah Schäfer-Althaus, Uta Schaffers (New York: Routledge, 2023), 48.

Michael Meyer contributed in the same volume with "Mary Wollstonecraft and the Body of Her Letters, or The Traveler Lost and Found in Scandinavia". Meyer discusses previous studies on Wollstonecraft as a gendered and embodied observer or working with aspects of the embodied self. Meyer focuses on "processes of embodied traveling, including the ominous beginning and ending of Wollstonecraft's journey omitted from the travelogue."[10] He stresses in his analysis that the narrator in the *Letters* can be regarded as an embodied self that is suffering but that is also responsible for the alienation from herself and the societies she is visiting.

Luisa Simonutti added a few notes on Mary Wollstonecraft in the chapter "Elsewhere. Women Translators and Travellers in Europe and the Mediterranean Basin in the Age of Enlightenment." Simonutti points also at the various roles of women travellers. They "played a myriad of roles as wives, mothers, lovers, adventuresses, slave traders, writers, philosophers, and scientists."[11] Simonutti places Wollstonecraft in the context of the eighteenth century in which more women started to travel and mediate ideas, cultures and customs.

To summarise, these recent studies on Wollstonecraft are closely connected to the genre of travel literature, the role of the women traveller and the place of the body (gender studies) and cultural transfer. It shows that the protagonist Wollstonecraft with her *Letters* has become the icon of travel writing. Scholars time and again find new perspectives and develop new approaches. In this chapter I will demonstrate that the last word has not yet been written and that the combination of persona and performativity with the approach of cultural transfer expands our understanding of the *Letters.*

The increasing interest in Wollstonecraft's *Letters* and the studies of travel literature are boosting each other. The study in travel writing became an important direction within cultural, historical and literary studies in the last decades. As a genre, travel writing has a long history that has accompanied and been intertwined with other genres, such as historical accounts and the novel. People travel out of curiosity (How do people live elsewhere? What is beyond the horizon?), migration (Where can I find a new home and food? Where am I safe and can survive?) and to conquer and colonise (Where can I find new territories to expand

10. Meyer, Michael. "Mary Wollstonecraft and the Body of Her Letters, or The Traveler Lost and Found in Scandinavia," in *Traveling Bodies. Interdisciplinary Perspectives on Traveling as an Embodied Practice*, eds. Nicole Maruo-Schröder, Sarah Schäfer-Althaus, Uta Schaffers (New York: Routledge, 2023), 85.

11. Luisa Simonutti, "Elsewhere. Women Translators and Travellers in Europe and the Mediterranean Basin in the Age of Enlightenment," in *Gender and Cultural Mediation in the Long Eighteenth Century. Women across Borders.* Mónica Bolufer, Laura Guinnot-Ferri, Carolina Blutrach (Eds.), 2024, 194.

my power and wealth?), to list a few reasons.[12] Travellers have written about their experiences and adventures for themselves, for those who stayed at home or even to report to authorities. The development of travel writing as a genre is thus closely connected to the people who were travelling and took up the pen to tell stories about their experiences during or after their voyage.

In 1795, Mary Wollstonecraft travelled to Scandinavia to help her former lover, Gilbert Imlay (1754–1828), investigate the disappearance of a ship. She reported her experiences and findings to him in a series of letters, and later decided to publish these.[13] Here, I will position Wollstonecraft's *Letters written during a short residence in Sweden, Norway and Denmark*, within the genre of travel writing, arguing that her writing about the art of travelling and travel writing can be regarded as a performative act. Consequently, I will reflect on the concepts of performativity and persona. The analysis of the *Letters* will focus on Wollstonecraft as a traveller-cultural transmitter and how she mediated ethnotypes of the Scandinavian countries she visited.

The perspective of cultural transfer in the *Letters* has not yet been investigated in any depth. In combining performativity and persona with the cultural transfer of ethnotypes, I aim to contribute to the knowledge about how cultural transfer functions as a performative act in travel writing. Wollstonecraft's ambition was to describe "national characters" in her *Letters* and, in this way, she acted as an ethnographer ahead of her time.

Wollstonecraft's *Letters* in the context of the history of travel writing

Percy G. Adams considers, with reference to L. Casson, that the ancient Greek writer and scholar Herodotus (c. 485–425 BC) was one of the first travel writers.[14] Later, the journeys of the Vikings to North America and other places in the period 800–1050 were described in accounts by Hauksbók (c. 1310) and Flateyjarbók (c. 1375). From the period these travel accounts were published, there are other more well-known names, such as the Venetian, Marco Polo (1254–1324), and Ibn Batuta (1304–1368/1369) from Morocco. More than half a century before, in 1253 and 1254, the Flemish monk, William of Rubruck, a missionary and explorer, travelled to Central Asia and Mongolia. These trips might be put in the

12. See also Barbara Korte, "Practices and Purposes," in *Handbook of British Travel Writing*, ed. Barbara Schaff (Berlin/Boston: De Gruyter, 2020), 95–112.

13. Mary Wollstonecraft, William Godwin, *A Short Residence in Sweden* and *Memories of the Author of 'The Rights of Women'*, ed. Richard Holmes (Penguin, 1987).

14. Percy G. Adams, *Travel Literature and the Evolution of the Novel* (Lexington, 1983), 45–6.

context of discovery, missionary work, colonisation and trading. Later, the Grand Tour for educational purposes for upper class young people, as well as science and tourism, became important reasons to travel, while the phenomenon of the professional traveller also became visible in the nineteenth and twentieth centuries.[15]

Travel writing thus had different forms and aims. Often the travellers aimed to provide information and transmit their experiences, emotions and impressions. These texts could be regarded as a mix of "faction" and fiction. We might also think of travel writing as a hybrid genre – faction with a poetic veil.[16] In the nineteenth century, travelling and thus writing about it expanded, as Barbara Schaff has demonstrated.[17] While Wollstonecraft travelled to Scandinavia at the end of the eighteenth century, I argue that her *Letters* can be regarded as a threshold in the history of travel writing. With their romantic elements imbued with a personal style of writing, the work connects to contemporary views on travelogues. Richard Holmes calls the *Letters* not only "one of the neglected masterpieces of early English Romanticism," but also a book that was the inspiration for travel writers later on, for example, Shelley and his wife, Mary Wollstonecraft's daughter Mary, as well as Robert Louis Stevenson (see also articles by van Tilburg and Sandrock in this volume).[18]

Holmes claims that Samuel Taylor Coleridge (1772–1834), when writing *Kubla Khan, or a Vision in a Dream. A Fragment* might have had the *Letters* in mind.[19] According to Holmes, Wollstonecraft was "a serious commercial traveller."[20] Because of the "combination of progressive social views" and "melancholy self-reve-

15. In addition to the material provided in later studies, such as Barbara Schaff, ed., *Handbook of British Travel Writing* (Berlin/Boston: De Gruyter, 2020), and Carina Lidström, *Berättare på resa. Svenska resenärers reseberättelser 1667–1829* (Stockholm: Carlson, 2015), this historical overview is based on a previous article that I published in Swedish in 1993: Petra Broomans, "Nordenbilden i några litterära reseskildringar," in *Skandinaviensbilleder. En antologi fra en europaeisk kulturkonference*, ed. Godelieve Laureys, Niels Kayser Nielsenand Johs. Nørregaard Frandsen, *Tijdschrift voor Skandinavistiek* 14, no. 1 (1993): 147–72. In this article, I also discussed Wollstonecraft's *Letters* in the context of the development of travel writing. In 2004, I published a short essay on Mary Wollstonecraft's *Letters* in the Netherlands, a first exploration of its reception in a small-language area.

16. Broomans, "Nordenbilden," 170.

17. Schaff, *Handbook*, 1.

18. Richard Holmes, "Introduction," in Mary Wollstonecraft, William Godwin, *A Short Residence in Sweden* and *Memories of the Author of 'The Rights of Women'*, ed. Richard Holmes (Penguin, 1987), 36.

19. Holmes, "Introduction," 41.

20. Holmes, "Introduction," 21.

lation and heart-searching," Wollstonecraft had "almost a symbolic force" within the group of "poets, travellers, philosophers and biographers" at that time.[21] Holmes sees her as a "model of the literary woman," who was forgotten in England during the nineteenth century, although "rare women traveller-writers as Isabella Bird and Mary Kingsley," did refer to Wollstonecraft.[22] Wollstonecraft herself was also likely influenced by previous travel writers. According to Leask, Wollstonecraft had reviewed George Forster's book *Voyage round the World* from 1777 and might have read his thoughts about 'the traveller's sensibility'.[23]

Having in mind that the novel was only developing as an important genre in the eighteenth century, we could also position Wollstonecraft's *Letters* against the backdrop of the shaping of the novel. According to P.G. Adams, travel writing was one of the accompanying genres in the development of the novel. At the end of his study, *Travel Literature and the Evolution of the Novel* (1983), Adams makes some interesting observations that still echo in later studies on travel writing. He commented, for example, that it is difficult to give a good definition of the genre of travel writing:

> The récit de voyage is not just a first-person journal or diary. Much of it is in third person. It is not just in prose. There are many poetic travel accounts ... and travel literature is often beautifully illustrated. ... It is not just a set of notes jotted down every day or whenever the traveler has time..., but far more often the account has been reworked, edited ... a fact that for the twentieth century is perhaps even more true.[24]

Adams also writes that the travel account is subjective rather than objective and can never be complete, and as he observes at the time of writing in the 1980s, few scholars had dealt with travel writing. Currently, 40 years later, travel writing as a genre has been explored by many scholars within the field of cultural studies and history, as well as in the fields of cultural transfer studies, imagology studies and mobility studies, as seen in the studies discussed above.

Another recent study on travel writing, edited by Barbara Schaff, is the *Handbook of British Travel Writing* (2020). Even if the volume deals specifically with British travel writing, the introductory chapters are fruitful for the research field in general. She refers to Tim Youngs, who proposed a definition of the travel genre in the *Cambridge Introduction to Travel Writing* (2013) and who also commented on the difficulty of providing a useful and decisive definition. As Schaff comments in

21. Holmes, "Introduction," 41.

22. Holmes, "Introduction," 41.

23. Leask, "Eighteenth-Century Travel Writing", 105.

24. Adams, *Travel Literature*, 280.

this regard: "Travel writing is a mercurial category, involving, and sometimes combining, elements of autobiography, memoir, diary, letters, journalistic reportage, fiction, poetry, satire, or travel guide."[25] This observation reflects the description given by Adams. Schaff also mentions the hybridity of the genre and an important observation by Barbara Korte that the process of writing the travel account moves "from fact to fiction."[26] The travel writer will need to fill in the gaps in the story when reconstructing the journey and will use fiction to do so. According to Schaff, the field of cultural studies and cultural translation includes travel writing as an object of study as well. We can add that travel literature has always been important for cultural transfer studies. Both Peter Burke and Michel Espagne regard travellers, merchants and missionaries, as important mediators of culture, literature and knowledge. Other fields mentioned by Schaff are, quite rightly, mobility studies, postcolonialism and gender studies.

Schaff states that travel writing "was shaped by the British."[27] While this is true in terms of the quantity of material, the history of travel writing, as depicted very briefly in the introduction, shows that through the ages other nations have also had a rich travel writing history. We could add to the list Dutch journeys around the Arctic region in search of the northeast passage to China, as Marijke Spies describes in her study, *Bij Noorden om. Olivier Brunel en de doorvaart naar China en Cathay in de zestiende eeuw* (1994). The mariners, merchants and diplomats used travel journals to report variously on the sea routes they discovered, the goods they bought and sold and the foreign countries they visited: "The travel journal is a genre that has flowered in the Netherlands since the late Middle Ages. Such documents form an important source both for Dutch history and that of the many countries visited."[28] Recently, Tonio Andrade published *The Last Embassy: The Dutch Mission of 1795 and the Forgotten History of Western Encounters with China* (2021), which is about the journey of Dutch diplomats to Beijing in 1794–1795.[29] One of the travellers he describes is Andreas Everardus van Braam Houckgeest (1739–1801), whose travel report was published in 2021 in Dutch translation (it was originally published in French in 1797/98) and can be compared to Wollstonecraft's *Letters*. The journey not only had a diplomatic pur-

25. Schaff, *Handbook*, 1.

26. Schaff, *Handbook*, 2.

27. Schaff, *Handbook*, 3.

28. R.M. Dekker, "Dutch Travel Journals from the Sixteenth to the Early Nineteenth Centuries," in *Lias. Sources and Documents relating to the Early Modern History of Ideas* 22 (1995): 277–300.

29. Tonio Andrade, *The Last Embassy: The Dutch Mission of 1795 and the Forgotten History of Western Encounters with China* (Princeton and Oxford: Princeton University Press, 2021).

pose, but also pursued commercial interests, which confirms that such journeys could be for several reasons.

At the beginning of the long nineteenth century, Dutch travellers explored Scandinavia and the Baltic countries. For example, the diplomat, Johannes Meerman (1753–1815), made a Grand Tour in his youth to England, France, Switzerland, Italy, Germany and Austria. Later, due to political problems, he made a long journey with his wife between 1797 and 1800 to Denmark, Sweden, Russia, Poland and Prussia.[30] Thus, Wollstonecraft had contemporaries who travelled to the North.

In her historical overview of travel writing, Schaff demonstrates that, in the competition between Britain, France and the Dutch Republic for "supremacy in the newly discovered parts of the world for most of the eighteenth century, Britain would become the dominant force."[31] This was already the case when three Dutch diplomats made their journey to Beijing in 1794–1795, the same period when Wollstonecraft made her Scandinavia journey.

In the chapter, "Discourse of Travel Writing" in the *Handbook of British Travel Writing* (2020), Sandrock discusses various types of travel writing throughout the ages, from "the scientific and the sentimental type"[32] to the empirical scientific modes in the eighteenth and nineteenth centuries.[33] Truth was important in scientific travel writing because the reader had to be certain that the information was correct. One good example given by Sandrock is Charles Darwin's *The Voyage of the Beagle* (1839).[34] In the twentieth century, the "truth" in travel writing was questioned because of its often subjective narrator and against the backdrop of expanding tourism and the quest to find the authentic: "Instead of an empirical or singular conception of truth, the self emerges as an organizing principle positioned between the text and the world."[35] Thus, according to Youngs,[36] the "narrative persona" becomes important.

---

30. Amy van Marken, "Johannes Meerman. A Dutch traveller to the North and North-East of Europe (1797–1800)," in *Baltic Affairs. Relations between the Netherlands and North-Eastern Europe 1500–1800*, ed. J.Ph.S. Lemmink and J.S.A.M. van Koningsbrugge (Nijmegen: INOS 1990), 229–48.

31. Schaff, *Handbook*, 18.

32. Referring to Casey Blanton 1997, 11, cited in Kirsten Sandrock, "Discourses of Travel Writing," in *Handbook of British Travel Writing*, ed. Barbara Schaff (Berlin/Boston: De Gruyter, 2020), 33.

33. Sandrock, "Discourses," 34.

34. Sandrock, "Discourses," 34.

35. Sandrock, "Discourses," 38.

36. Cited in Sandrock, "Discourses," 38.

Another interesting reflection is that of Christoph Bode, who considers the self and how it developed in the Romantic period "towards a highly subjective, self-reflective discourse of the travelling persona."[37] Here, Sandrock also refers to Mary Wollstonecraft's travel account and states that it also reflects a "turn towards self-reflectivity."[38] This self-reflexive mode has also been observed by Richard Holmes in his introduction to Wollstonecraft's travel book.[39]

To summarise, Mary Wollstonecraft's *Letters* are a natural part of the development of travel writing and the role of women in this development at the end of the eighteenth century. The *Letters* also mark a threshold in the history of travel writing by combining commerce, facts and emotions as well as a poetic veil. The *Letters* are also perfect material for defining the genre. In the centre is the I-narrator, Mary Wollstonecraft, who adopts different roles in the work, as already noted by Radu, Zold and Simonutti.

Performativity and persona

In his introduction, Holmes observes that Wollstonecraft showed "the extraordinary skill with which she transformed a prosaic business venture into a poetic revelation of her character and philosophy."[40] Character and agency are important features in Wollstonecraft's *Letters*. Agency can be linked to performance and character to the concept of *persona*. Performativity is observed, although not labelled as such by Holmes. When writing about Letter Nineteen, for example, he contends that it "is her most explicitly feminist chapter."[41] In this way, Holmes attributes both agency and performance to Wollstonecraft. Furthermore, insofar as Wollstonecraft wrote *A Vindication of the Rights of Women* with the aim to change the world for women, this is also an important example of performativity.

Performativity is a term that developed in the field of philosophy, and it was modified as well as applied in the fields of linguistics, sociology and gender. Jillian R. Cavanaugh, inspired by John L. Austin, states that performativity is closely linked to language; it is "the power of language to effect change in the world."[42] Performativity is in fact the process of subject formation by linguistic means and

37. Cited in Sandrock, "Discourses," 38–39.

38. Sandrock, "Discourses," 39.

39. Holmes, "Introduction," 32.

40. Holmes, "Introduction," 26.

41. Holmes, "Introduction," 31.

42. Jillian R. Cavanaugh in *Oxford Bibliographies*. See: https://www.oxfordbibliographies.com/view/document/obo-9780199766567/obo-9780199766567-0114.xml

social practices. It can also refer to the desire to change the world. In the field of gender studies, Judith Butler elaborates Austin's notion of performativity in *Gender Trouble* (1990):

> According to Austin, in order for a statement to have performative force (in other words, in order to enact what it names), it must (1) be uttered by the person designated to do so in an appropriate context; (2) adhere to certain conventions; and (3) take the intention(s) of the utterer into account.[43]

Additionally, all linguistic signs are vulnerable to appropriation. The following statements show the influence of Simone de Beauvoir (1908–1986), who wrote in her seminal book, *The Second Sex* (1949), that one is not born but becomes a woman, and thus identity is a construction. Judith Butler elaborated on De Beauvoir's thoughts, introducing the term "gender." She argues that gender identities are constructed and constituted by language: language and discourse *do* gender. Thus, gender identity is performative and *political*, in the sense of the feminist slogan, "the personal is political," referring to the fact that power is constitutive of our relationships.

Within speech act theory, a performative act is the discursive practice that enacts or produces that which it names. Thus, linguistic constructions create our reality through the speech acts we participate in every day. In the performative act of speaking, we incorporate reality and make conventions appear natural. A speech act is a bodily act; there must be a body that speaks and which thus exists in discourse. What is important in performativity is not simply the act that is performed, but primarily the fact that the act is done repeatedly. To say that gender is performative means that nobody *is* a gender prior to *doing* gender acts: gender only exists in being performed: gender acts and gender identity exist at the same time.

What role does the travel writer, in this case, Mary Wollstonecraft, want to perform? As we shall demonstrate, Wollstonecraft acted as a cultural transmitter of ethnotypes, acted as a scholar, wanted to convince Imlay to take her back as his lover by acting as his investigator, took the role of philosopher, politician and observer of gender roles, and she also showed an awareness of what she was doing: writing travel literature and reflecting on the genre. She is aware of the blurring of memory, in order to describe in a correct way what she had seen during the travels. Wollstonecraft had to relate to the "effect different objects had produced on my mind and feelings, whilst the impression was still fresh" (*Letters*: 62).

43. Quoted in Sara Salih, "On Judith Butler and Performativity," in Sara Salih, *Judith Butler*. Routledge Critical Thinkers (London & New York: Routledge, 2002), 62.

These aims and positions can be linked to performativity because every role requires a specific speech and voice. Wollstonecraft's *Letters* are thus a fine example of performativity and gender identity, with the writing of the letters during the journey and their editing in another place both performative acts. These acts resulted in the *Letters* that are still read today; we hear the author's voice still speaking through the text, which might be regarded as symbolic of the body.

Wollstonecraft's own personal reflections on travel writing and travelling, as well as her thoughts on politics and being a woman, also led me to the concept of the persona and the question of the connection between performativity and persona. Carina Lidström's study on Swedish travel writing between 1667 and 1829 discusses the term "persona" and how it can be used in the study of travel writing.[44] Lidström's study not only offers rich material on Swedish travel writing but also interesting theoretical insights regarding genre definitions, repertoires and contracts. Lidström also describes the genre of travel writing as a hybrid and fluid genre.

It is important to consider the context in which the text is written, as well as its function, by examining for example whether clear instructions were given to the traveller by someone else or whether the latter expected trustworthy information. In this case, the travel account can be considered as an informative text rather than a subjective narration of the traveller. If the travel was purely for pleasure, readers would expect another kind of text. This can be determined by what Lidström calls, referring to Lejeune, a "genre contract."[45] A good example of functional travel texts is the category of the "periploi" or guides for travellers.[46] Texts related to this category fulfil the wish to know "how life is over there," they are a kind of in-between text, partly a genre contract of true facts while also satisfying the curiosity of the readers. Lidström also works with Jauss's concept of "Erwartungshorizont,"[47] and addresses the perspective of mediation, which will also be investigated below.

Regarding the travel narrator, Lidström applies Lejeune's definition of the genre of autobiography to travel writing with respect to the narrator – author, narrator and protagonist are one and the same person.[48] Thus, travel literature is an autobiographical genre that enables the travel writer to take the role of their own

---

44. Carina Lidström, *Berättare på resa. Svenska resenärers reseberättelser 1667–1829* (Stockholm: Carlson, 2015).

45. Lidström, *Berättare*, 23.

46. Lidström, *Berättare*, 23.

47. Lidström, *Berättare*, 30.

48. Lidström, *Berättare*, 31.

choice. The basic role of the travel writer is of course to be "the traveller," but there are several additional roles. It is here that Lidström uses the term "persona."[49]

Lidström discusses the origin and different meanings of the term "persona," referring to the Latin word for persona that means "mask" and to the "dramatis personae" in classical drama.[50] Lidström also refers to C.G. Jung's theories about the different roles a person can take, suggesting that a persona does not necessarily refer to the whole personality but only to a part of it. In relation to travel writing, Lidström uses the term "persona" to emphasise that the role of the traveller is taken both within the travel account and in reality.[51] Here, the persona is determined by both social context and literary conventions. According to Lidström, the role of travel narrator could also be described on the basis of Bourdieu's concept of "habitus."[52] The repertoire of a person's agency and self-presentation in the travel account can thus be contextualised and linked to the goal of the travel writer.

Examples of persona in travel writing include the adventurer, the reporter, "our man on the spot,"[53] scholars and aesthetes. The persona in travel writing often uses paratexts,[54] which we will also observe in the *Letters* of Wollstonecraft. The "Advertisement" (*Letters*: 62) is an example of a specific type of paratext called a peritext. The focus on taking on a role and choosing a voice in travel writing is a good example of how the notion of performativity can be added to the persona. This will be demonstrated in the analysis, where performativity will be applied to the persona in travel writing and to the kind of persona encountered in Mary Wollstonecraft's *Letters*.

It becomes clear that Mary Wollstonecraft had various aims with her *Letters*, such as reporting on her investigation into the lost ship, solving the mystery and tempting Imlay to take her back as his partner. These tell us about the social and mercantile context of her *Letters*, as well as the emotional goal. A third goal of the *Letters* is to inform readers about what she considers to be the national character

49. Lidström, *Berättare*, 35.

50. Lidström, *Berättare*, 35.

51. Lidström, *Berättare*, 35.

52. Lidström, *Berättare*, 37.

53. Lidström, *Berättare*, 74.

54. The term "paratext" was coined by Genette. It concerns texts by authors that are separate from the literary text but are linked to the work. Paratexts offer the reader more knowledge and understanding about the text. Paratexts can be divided into peritexts and epitexts. According to Genette, a foreword is a peritext. Interviews with the author and lectures by the author are epitexts, G. Genette, *Paratexts: Thresholds of Interpretation*, trans. Jane E. Lewin (Cambridge: Cambridge University Press, 1997), 2.

of the Danes, the Norwegians and Swedes. In describing these national characters, Wollstonecraft acts here as an anthropologist, and in informing the reader about these national characters she also operates as a cultural transmitter. Within the field of imagology, the term "ethnotype" is used for descriptions of national character. In this article, the former will be used on the meta-level (of academic discourse), while the latter (traditions, customs, emotions) will refer to the practice of doing and presenting the national character in real life.

Ethnotypes and cultural transfer

The focus on the cultural transfer of ethnotypes is based on the fact that this was actually one of the aims of Mary Wollstonecraft: "My plan was simply to endeavour to give a just view of the present state of the countries I have passed through" (*Letters*: 62) and further motivated by her interest in describing the national character of the Swedes, the Norwegians and the Danes. We have to bear in mind that the "just" view is what Wollstonecraft sees as just and that her view is coloured by her own political and ideological lens. Moreover, she informs her readers about her goals in the "Advertisement." Ethnotypes in travel writing have especially been analysed in the field of imagology:

> The resulting "ethnotypes" (frivolous French, passionate Hungarians, thoroughgoing Germans) are commonplaces which change over time and migrate between genres and countries. While they have little or no empirical basis, they do exercise a real influence in how people judge the world and their position in it. Ethnotypes are both a complex culture-historical phenomenon and an important part of nationalistic prejudice and ethnocentrism, and stand in need of critical analysis for both reasons.[55]

Ethnotypes are also transferred in the cultural transfer process when a cultural object is transferred from a source context into a receiving context.[56] In this transfer process, the cultural transmitter can take on different roles, such as translator, bookseller and/or traveller. In travel writing, information about culture in the broadest sense of the word (history, politics, traditions, people's social behaviour) as well as information about the literary field (authors, literary works) is transmitted. To give an example: Dutch authors travelled in Scandinavia in the nineteenth

55. See https://imagologica.eu/.

56. Michel Espagne, "Jenseits der Komparatistik. Zur Methode der Erforschung von Kulturtransfer," in *Europäische Kulturzeitschriften um 1900 als Medien transnationale rund transdisziplinärer Wahrnehmung*, ed. U. Mölk (Göttingen: Vandenhoeck & Ruprecht, 2006), 15.

century and commented on foreign literature for Dutch readers.[57] These writers had similar mercantile purposes for their journeys to those of Wollstonecraft, but they also acted as cultural transmitters and, as Lidström's notion of "persona" suggests, they played different roles.

Thus, Wollstonecraft's *Letters* can be approached from a number of different perspectives. In addition to the cultural transfer perspective, there is also the related discipline of mobility studies, also mentioned by Schaff. Mobility studies deals with "the hidden and conspicuous movements of peoples [*travellers*], objects [*books*], images [*national characters, ethnotypes*], texts [*travel literature*], and ideas [*ideologies*]."[58] (See also the article by Eduardo Callegos Krause in this volume). The analysis of the *Letters* below, will explicitly focus on travel writing as cultural transfer, performativity, persona and ethnotypes. These angles will be developed in the analysis of the *Letters*, focusing on the way in which Wollstonecraft writes about travel writing and about mediating ethnotypes of the countries she visited.

Before analysing the *Letters* in depth, I will give a short description of the text. The book consists of 25 letters that cover the three and a half months that Wollstonecraft spent in Scandinavia and were originally written to Gilbert Imlay (1754–1828), her former lover and father of her daughter Fanny. The purpose of the journey was trade and to act as an investigator on behalf of Imlay, who had lost a ship in Scandinavia. Wollstonecraft also hoped to regain the lost love of Imlay. Wollstonecraft took her daughter Fanny, then one year old, and her French maid Marguerite with her. In Norway, she left Fanny and Marguerite in Gothenburg in the home of Elias Backman, Imlay's Swedish agent (*Letters*: 282). When she had to leave her little daughter in the care of Marguerite for the first time, she describes her feelings as a mother and expresses concerns about her daughter's future as a girl and as a woman.[59] Even during her journey, Wollstonecraft had already decided to publish the letters, and they appeared in 1796, published by William Godwin. This, then, is also relevant to our understanding of her relation to the text. Holmes characterised Wollstonecraft as "a born traveller and an

57. See Petra Broomans, "Alone! Alone!," in *The Swedes & the Dutch were made for each other*, ed. Kristian Gerner (Lund: Historiska Media, 2014), 180–91.

58. Stephen Greenblatt, "A mobility studies manifesto," in Stephen Greenblatt et al., *Cultural Mobility. A Manifesto* (Cambridge: Cambridge University Press, 2009), 250–3.

59. Laurie Langbauer has written an interesting chapter about motherhood in Wollstonecraft's novels and made some comparisons with *The Vindication of the Rights of Woman*. Laurie Langbauer, *Women and Romance. The Consolations of Gender in the English Novel* (New York: Cornell University Press, 1990). As mentioned, in a recent study by Zold, the importance of her daughter during this episode is described in an empathetic way.

instinctive seeker after new horizons and new societies" (*Letters*: 22), and it is not surprising that Wollstonecraft, as a woman of letters and aware of her aims, also writes about the art of travelling and travel writing, which can be regarded as a performative act.[60]

Mary Wollstonecraft by John Opie (c. 1797)

60. Wollstonecraft died one year after the *Letters* were published, shortly after childbirth. Her daughter Mary married Percy Shelley and became the famous author of *Frankenstein. The Modern Prometheus* (1818, published anonymously. In 1831 she published a revised version under her own name).

Wollstonecraft on travel writing and travelling

For Wollstonecraft, travel should be "a kind of sociological inquiry, which is a much more critical and comparative idea of how societies develop and progress."[61] As Holmes states, referring to Letter Nineteen, Wollstonecraft's attitude "foreshadows much of the more strictly anthropological travelling of the nineteenth century, with its emphasis on comparative studies of particular societies and climates."[62] For that matter, in these sections, Wollstonecraft manifests herself as academic, which, in the context of the position of learned women in her time, is an important role for Wollstonecraft.

In the "Advertisement" (*Letters*: 62), in which Mary Wollstonecraft explains her goals, she also writes that she could not avoid being "the little hero of each tale," but that she tried to correct this "fault" because she intended to publish the letters. Nevertheless, she felt that her writing became "stiff and affected." She notes that she chose to let her comments, thoughts and reflections "flow unrestrained," as a kind of stream of consciousness, and adds that it was impossible to describe everything correctly and that she could not do otherwise but relate "the effect different objects had produced on my mind and feelings, whilst the impression was still fresh" (*Letters*: 62). This shows that she was very aware of her position as a travel writer and the difficulties of such writing. Furthermore, Wollstonecraft shows theoretical insight and an awareness of the impossibility of writing a true account of the journey. She reflects on the value of travel, which she considers can be regarded as a "completion of a liberal education," and advises travellers to visit the Nordic countries before travelling to "the more polished parts of Europe." She also warns readers not to equate hospitality with the "virtues of a nation; which, I am now convinced, bear an exact proportion to their scientific improvements" (*Letters*: 173).

Even if Wollstonecraft belongs to a new generation of women travel writers, she prefers to be regarded as equal to a male travel writer. She did not only describe households and clothing, but according to Leask, she was proud to be regarded as an observer who also raised "men's questions".[63] In her letters, Wollstonecraft also mediates her images of men and women and relates what it is like to travel alone as a woman, revealing that this both interested the people she met and also that they "wished to protect me" (*Letters*: 96/97). Furthermore, acting as an investigator and business woman in a male-dominated domain also draws attention.

In the Appendix, Wollstonecraft adds information about the preparation of the letters for publication:

61. Holmes, "Introduction," 32.

62. Holmes, "Introduction," 33.

63. Cited in Leask, "Eighteenth-Century Travel Writing", 105.

> … as a person of any thought naturally considers the history of a strange country to contrast the former with the present state of its manners, a conviction of the increasing knowledge and happiness of the kingdoms I passed through, was perpetually the result of my comparative reflections.
>
> The poverty of the poor, in Sweden, renders the civilization very partial; and slavery has retarded the improvement of every class in Denmark; yet both are advancing; and the gigantic evils of despotism and anarchy have in a great measure vanished before the meliorating manners of Europe. (*Letters*: 198)

In this peritext, Wollstonecraft evaluates ethnotypes she observed in the Scandinavian countries she visited. I will now turn to the question of how she writes on ethnotypes as an early form of anthropology and as manifesting her academic mindset.

Wollstonecraft on national characters

Wollstonecraft is convinced that one can distinguish "the manners of a people" better in the country, the rural regions, than in the cities (*Letters*: 85), where people are everywhere "the same genus" (*Letters*: 85). In observing that the people in Norway were different from people in Sweden, she adds that travel writers "are eager to give a national character" (*Letters*: 92) and criticises writers who expect that the countries they visit should resemble their own. They "had better stay at home" (*Letters*: 93), she suggests. In Letter Six, she remarks that while she does not understand the Norwegian language (*Letters*: 99), she thinks that she was able to gain an impression of "the character of the Norwegians, without being able to hold converse with them" (*Letters*: 113).

Wollstonecraft regarded Norway as revolutionary and free, despite it still being part of the Danish monarchy at the time: "Though the king of Denmark be an absolute monarch, yet the Norwegians appear to enjoy all the blessings of freedom" (*Letters*: 101), and she remarks that although the Norwegians have more connection with the English, they also sympathise with the French Revolution (*Letters*: 140).

When Wollstonecraft leaves Gothenburg, she laments not being able to see more of Sweden, but she had already formed some impressions about the country:

> … yet, I imagine I should only have seen a romantic country thinly inhabited, and these inhabitants struggling with poverty. The Norwegian peasantry, mostly independent, have a rough kind of frankness in their manner; but the Swedish, rendered more abject by misery, have a degree of politeness in their address, which, though it may sometimes border our insincerity, is oftener the effect of a broken spirit, rather softened than degraded by wretchedness. (*Letters*: 160)

Wollstonecraft adds further remarks about the poverty of the Swedes (*Letters*: 161). Regarding family life and marriage, Wollstonecraft comments that the Swedes are "attached to their families" and divorce can be obtained when there is proof of infidelity. Reflecting on this, she writes that affection "requires a firmer foundation than sympathy" (*Letters*: 161), but admits that she is moralising, revealing once again her self-awareness as a women travel writer.

In Denmark, Wollstonecraft describes the business men as "domestic tyrants, coldly immersed in their own affairs," and that they believe Denmark to be the happiest country in the world, which is doubted by Wollstonecraft (*Letters*: 165). In her depiction of the women in Denmark, she is aware that she might be "a little prejudiced, as I write from the impression of the moment" (*Letters*: 165). She regards the women as "simply notable house-wives; without accomplishments, or any of the charms that adorn more advanced social life," and she did not see them as good parents (*Letters*: 165). That she regards herself as prejudiced also has to do with the case of Queen Matilda (1751–1775), the sister of the English king, George III (*Letters*: 165).[64] She writes that she was made angry "by some invectives thrown out against the maternal character of the unfortunate Matilda" (*Letters*: 165). Wollstonecraft, herself British, describes how Matilda raised her children in a natural way, as advocated by Rousseau. Wollstonecraft thus presents Matilda as a better parent than the Danish women who criticised her. She also defends the goals and works of Matilda and the royal physician Johann Friedrich Struensee (1737–1772), who wanted to install a liberal government, criticising the king, who was "an idiot into the bargain" (*Letters*: 166). Struensee, who was probably the queen's lover, was beheaded, and Matilda was sent to Celle, where she died at the age of 24.

Wollstonecraft probably sympathised with Matilda and Struensee because of her own political preferences. She writes that Matilda "was the victim of the party she displaced," but also that both Matilda and Struensee wanted to carry out the reforms too quickly, "she probably ran into an error common to innovators, in wishing to do immediately what can only be done by time" (*Letters*: 166). The case of Matilda accords with Wollstonecraft's own concern with the ideals of emancipation and social reform and also demonstrates that women could be innovators in the domestic and in the public spheres, though frequently hindered by society.[65]

In Letter Twenty, Wollstonecraft informs the reader about the theatre, literature, arts and science in Denmark. She has to admit that the public library in Copenhagen has "a collection much larger than I expected to see" (*Letters*: 176), including famous Icelandic manuscripts. In the royal museum she notes there are

64. Cf. Holmes, "Introduction," 289, n. 104.

65. I thank Kirsten Sandrock for this observation.

"some good pictures," while a "collection of the dresses, arms, and implements of the Laplanders" catches her attention. She describes the Sámi as the "first species of ingenuity which is rather a proof of patient perseverance, than comprehension of mind" (*Letters*: 176).

As far as a national character is concerned, she makes only general observations and states that she does not "pretend to sketch a national character; but merely to note the present state of morals and manners, as I trace the progress of the world's improvement" (*Letters*: 172). Thus, the ethnotypes Wollstonecraft transmitted reflect her own opinions on rural and urban poverty in Sweden and the free farmer in Norway. In Denmark, she visited the capital, Copenhagen, and discussed the culture in the capital, the different approaches to raising children and political circumstances.

Performativity and persona in practice

In Norway, Wollstonecraft's reflections on nature are often combined with observations of the climate. She mentions that the Dutch had come to Norway 50 years earlier and cut down trees in the forests and that the farmers were happy that they did not have to do it themselves. She comments: "The destruction, or gradual reduction, of their forests, will probably improve the climate; and their manners will naturally improve in the same ratio as industry requires ingenuity" (*Letters*: 121). This suggests that Wollstonecraft favours the development of industry, although she loves the pure air in Norway. Is it ambivalence or irony? The novels by Dickens about the disadvantages of polluting industry and bad working conditions in England, which challenged a belief in the benefits of progress, were yet to come.

Wollstonecraft also informs Imlay (and her readers) about her investigation of stories of monsters of the deep that were said to lurk in the northern sea. She reports having asked several seamen and captains, but none had seen them nor could give a real description of them. She concludes: "till the fact be better ascertained, I should think the account of them ought to be torn out of our Geographical Grammars" (*Letters*: 153).[66] Again, Wollstonecraft positions herself as a scholarly advisor here. As previously remarked, the scientific turn in travel writing that had already occurred, meant that there was motivation for travel writers to describe their adventures truthfully. Thus, Wollstonecraft's comment could be

66. Holmes comments on the "perennial interest in monsters of the deep" and regards Wollstonecraft's questions about real sightings as ironic; Holmes, "Introduction," 289, n. 104.

interpreted as an example of the turn towards scientific discourse in the genre of travel writing.

The next excerpt of the *Letters* demonstrates that Wollstonecraft also had an opinion about punishment and criminals.

> In fact, from what I saw, in the fortress of Norway, I am more and more convinced that the same energy of character, which renders a man a daring villain, would have rendered him useful to society, had that society been well organised. When a strong mind is not disciplined by cultivation, it is a sense of injustice that renders it unjust. (*Letters*: 168–169)

Wollstonecraft also regards the "adoration of property" as "the root of all evil" (*Letters*: 170). While Holmes regards her as an early anthropologist, in my opinion she could also be labelled as a sociologist. Wollstonecraft combines the study of human behaviour at the micro level, examining the way in which the maids are treated, relating traditions in eating and the clothing worn by individuals, with the study of power relations and politics. She discusses the political situation in each country she visited, and the relations between the Norwegians and the Danes in particular, to mention some topics at the macro level.

Trollhättekanal 1798 by Louis Belanger. Published 1800

Recalling her visit to the falls in the Swedish town of Trollhättan, near Gothenburg, Wollstonecraft returns to the theme of industry and nature. Sweden was building the Trollhätte Canal at that time, and Wollstonecraft experiences the "noise of human instruments and the bustle of workmen" who were working on the construction of the sluices (*Letters*: 160). Although apparently in favour of development and industry, she regrets that the landscape of the falls could not have been left "in all its solitary sublimity" (*Letters*: 160).

Performativity is included in her self-presentation as an early sociologist and observer of the implications of industry on the landscape. At the same time, Wollstonecraft believes that industry will contribute to the development of the people, using words such as "convinced" and "facts." Holmes has written a striking description of this performativity in the *Letters*.

> Mary Wollstonecraft projected herself through the book as a model of the literary woman: audacious, intelligent, independent and free-thinking; and yet, equally, one who suffers endlessly and inevitably in a society which is not yet honest and just enough to accept her for what she is.[67]

Radu refers to the same contradiction in the *Letters*, suggesting that while Wollstonecraft presents herself to Imlay as childlike and emotional, she also maintains her pride and self-confidence.[68] The persona in the *Letters* is thus complex and reveals different, even contradictory, traits. Addressing Imlay directly, she laments the loss of happiness that accompanies a childlike innocence and the honesty that she feels she has never lost. Stressing her feelings as a woman, she does not hesitate to use the words "grief" and "wounded":

> – for the grief is still fresh that stunned as well as wounded me – yet never did drops of anguish like these bedew the cheeks of infantine innocence – and why should they mine, that never were stained by a blush of guilt? Innocent and credulous as a child, why have I not the same happy thoughtlessness? (*Letters*: 189)

Elsewhere, she also reflects on this "extreme affection of my nature," again addressing Imlay (*Letters*: 111).

Another personal note can be found in Letter Twenty-Three, in which she addresses Imlay once more in an explicit way, after commenting on men's characters in general. She notes that priests are cunning, statesmen false and men of commerce make a "display of wealth without elegance, and a greedy enjoyment of pleasure without sentiment" (*Letters*: 191). Assuming that Imlay will consider her to be embittered, she implicitly seems to suggest that Imlay himself was a representative of men in general, and appears to conclude that for men, trade comes before everything:

67. Holmes, "Introduction," 41.

68. Radu, "Mary Wollstonecraft," 285.

> Men are strange machines; and their whole system of morality is in general held together by one grand principle, which loses its force the moment they allow themselves to break with impunity over the bounds which secured their self-respect. (*Letters*: 193)

Leask states that the format of letters "brilliantly balances the existential losses as well as the material gains involved in the modernisation of society, proposing a feminist solution to the corruptions attendant upon the commercial state, personified in her treacherous correspondent Imlay."[69]

Back in Hamburg, she compares the flat land in Germany with the mountains in Norway, where the rocks surrounded her, "shutting out the sorrow ... whilst peace appeared to steal along the lake to calm my bosom." In Hamburg, by contrast, she hears "only an account of the tricks of trade, or listens to the distressful tale of some victim of ambition" (*Letters*: 295). Clearly, she prefers the solitude of nature over the urban affairs of commerce. However, in earlier letters she appears to advocate for industry as a tool of development. Thus, Wollstonecraft expresses an ambivalence towards the value of commerce, industry and nature.

In Altona, near Hamburg,[70] she feels the need to change the scene, but reflects on this desire: "quitting people and places the moment they begin to interest me. – This also is vanity!" (*Letters*: 197). She appears tired of travelling and of meeting people she always has to leave too soon. Throughout the *Letters*, Wollstonecraft reflects on her own way of writing and the connection between emotion and objective observations:

> How often do my feelings produce ideas that remind me of the origin of many poetical fictions. In solitude, the imagination bodies forth its conceptions unrestrained, and stops enraptured to adore the beings of its own creation. These moments are of bliss; and the memory recalls them with delight. (*Letters*: 119)

For Wollstonecraft, feelings are thus a source of ideas and poetical fiction. Part of the persona in the *Letters* is a writer who sometimes feels blocked by her own thoughts. She feels that she cannot "write composedly – I am every instant sinking into reveries – my heart flutters, I know not why" (*Letters*: 136).

Nevertheless, she gains new inspiration from nature; and the journey is even a life-changing event. She remarks that it takes a long time "to know ourselves" and that in the solitude of nature in Norway, she has turned over "a new page in the history of my own heart" (*Letters*: 122). She experiences the solitude of her walks

69. Leask, "Eighteenth-Century Travel Writing", 106.

70. Altona is now part of Hamburg. It was under Danish control until 1864.

in the natural landscape as desirable, realising it awakens new ideas, without having a book with her (*Letters*: 132).

In the Appendix, again a peritext (*Letters*: 198) written during the preparation of the manuscript for the press, Wollstonecraft evaluates the *Letters*. She had subsequently become aware of what she had not described due to her investigation of what had happened to Imlay's ship and because she was hindered by personal concerns and affairs the entire time. She concludes with a brief socio-political analysis of the current state of affairs in the development of the Northern countries. While she sees improvements, evil, despotism and slavery still remain, and these "afflict the humane investigator" (*Letters*: 198) he warns against introducing improvements into society too quickly and altering "laws and governments prematurely" (*Letters*: 198). This process has to occur over time and cannot be "forced by an unnatural fermentation" (*Letters*: 198). It is possible that she is referring to the revolution that Queen Matilda and Struensee had wanted to start in Denmark and to her own experiences while she lived in Paris during the French Revolution. Leask mentions the catastrophic failure of the French Revolution and "the immediate backdrop to Wollstonecraft's travels."[71]

Wollstonecraft closes the Appendix with the following words:

> And, to convince me that such a change is gaining ground, with accelerating pace, the view I had of society during my northern journey, would have been sufficient, had I not previously considered the grand causes which combine to carry mankind forward, and diminish the sum of human misery. (*Letters*: 198)

In the Appendix, Wollstonecraft excels as a self-reflective observer and as a political scholar ahead of her time. She regards herself as a "humane investigator" and performs as such, also part of her persona in the *Letters*.

Conclusion

In the *Letters*, Wollstonecraft transfers to the readers her ideas on politics and the national characters of the people of Scandinavia (Denmark, Norway and Sweden), comparing the three countries. She is acting as an early anthropologist, as Holmes remarks, and as a sociologist. Wollstonecraft does not use these terms herself, but in the Appendix she reveals a self-awareness in stating that her findings are "the result of my comparative reflections" (*Letters*: 198). She also regrets that the commercial aim of her voyage often frustrated her desire to obtain information about the countries she visited (*Letters*: 198). She makes comments

71. Leask, "Eighteenth-Century Travel Writing", 106.

on both men's and women's roles. Imlay's attitude towards commerce seems to Wollstonecraft to be representative of men in general. As Holmes writes in his introduction, the *Letters* offer "Wollstonecraft's reflections on men and affairs,"[72] whereas women are depicted as housewives or we are introduced to a queen who is seen as an innovator in wanting to bring up her children in a modern way. There is an awareness of political and societal developments over time in this example of Mathilde and Struensee. Her observations of women are thus both on the micro level (clothing, raising children) and the macro level (politics, a queen who wanted to modernise Danish society).

According to Holmes, the *Letters* are "consciously literary in many aspects."[73] Indeed, as commented above, Wollstonecraft's *Letters* have a poetic veil and they brilliantly demonstrate the hybridity of the genre of travel writing. Even her writing on everyday life with her little daughter and maid, on meeting businessmen, is a convincing mix of fact and fiction. The more poetic parts deal with her impressions of nature as well as philosophical and aesthetic themes. The *Letters* are not naive travel writing for leisure or "societal/experience," nor "political" travel writing with a sole mediating purpose. As Holmes characterises in a splendid way, Wollstonecraft combines "progressive social views – Wollstonecraft's favourite subject of contemplation, the future improvement of the world – with melancholy self-revelation and heart-searching."[74]

Wollstonecraft constructs a new form of travel writing from a gendered perspective, and the *Letters* are revolutionary in the context of the time they were written. She shows a vulnerability in her writing about the woman traveller, revealing her emotions, and a self-awareness in admitting that she might be prejudiced and moralising.[75] Thus, Wollstonecraft's *Letters* not only mediate national characters, ethnotypes, but also her own political and societal ideas, her images of men and women. The persona in the *Letters* consists of her "self," the women travel writer, the grieving woman who has lost her lover, the mother that had to leave her little daughter and the confident scholar. By using language that expresses the different parts of her persona related to performativity, Wollstonecraft presents the persona as a revolutionary, though cautious, ultimate observer. Combining the cultural transfer of ethnotypes with persona (the self) and performativity (as writer, observer and scholar) in travel writing, the *Letters* demonstrate cultural transfer as a provocative performative act. In addition to the

72. Holmes, "Introduction," 35.

73. Holmes, "Introduction," 37.

74. Holmes, "Introduction," 41.

75. See also Elizabeth A. Bohls, "Gender," specifically the section "Gendered journeys," in *Handbook of British Travel Writing*, ed. Barbara Schaff (Berlin/Boston: De Gruyter, 2020), 61–3.

recent studies about Mary Wollstonecraft and the *Letters*, this chapter provided a more profound understanding of the multifaceted concept of cultural transfer through travel writing.

References

Adams, Percy G. *Travel Literature and the Evolution of the Novel.* Lexington, 1983.

Andrade, Tonio. *The Last Embassy: The Dutch Mission of 1795 and the Forgotten History of Western Encounters with China.* Princeton and Oxford: Princeton University Press, 2021.

Ayres, Brenda. *Betwixt and Between: The Biographies of Mary Wollstonecraft.* London, New York and Delhi: Anthem Press, 2017.

doi Bohls, Elizabeth A. "Gender." In *Handbook of British Travel Writing*, edited by Barbara Schaff, 55–78. Berlin/Boston: De Gruyter, 2020.

Broomans, Petra. "Alone! Alone!" In *The Swedes & the Dutch were made for each other*, edited by Kristian Gerner, 180–191. Lund: Historiska Media, 2014.

Broomans, Petra. "Nordenbilden i några litterära reseskildringar." In *Skandinaviensbilleder. En antologi fra en europaeisk kulturkonference*, edited by Godelieve Laureys, Niels Kayser Nielsen and Johs. Nørregaard Frandsen, *Tijdschrift voor Skandinavistiek*, 14, no. 1 (1993): 147–72.

Broomans, Petra. "Mary Wollstonecraft in Scandinavia; her letters in the Netherlands." In *"I have heard about you". Foreign Women's Writing Crossing the Dutch Border: from Sappho to Selma Lagerlöf*, edited by S. van Dijk, P. Broomans, J. van der Meulen, P. van Oostrum, 248–53. Hilversum: Verloren, 2004.

Cavanaugh, Jillian R. in *Oxford Bibliographies.* See: https://www.oxfordbibliographies.com/view/document/obo-9780199766567/obo-9780199766567-0114.xml

Dekker, R.M. "Dutch Travel Journals from the Sixteenth to the Early Nineteenth Centuries." In *Lias. Sources and Documents relating to the Early Modern History of Ideas* 22 (1995): 277–300.

Espagne, Michel. "Jenseits der Komparatistik. Zur Methode der Erforschung von Kulturtransfer." In *Europäische Kulturzeitschriften um 1900 als Medien transnationale rund transdisziplinärer Wahrnehmung*, edited by U. Mölk, 13–32. Göttingen: Vandenhoeck & Ruprecht, 2006.

doi Genette, G. *Paratexts: Thresholds of Interpretation.* Translated by Jane E. Lewin. Cambridge: Cambridge University Press, 1997.

doi Greenblatt, Stephen. "A mobility studies manifesto." In *Cultural Mobility. A Manifesto*, by Stephen Greenblatt et al., 250–3. Cambridge: Cambridge University Press, 2009.

Holmes, Richard. "Introduction." In *Wollstonecraft, Mary, William Godwin, A Short Residence in Sweden and Memories of the Author of 'The Rights of Women'*, edited by Richard Holmes. Penguin, 1987.

doi Korte, Barbara. "Practices and Purposes." In *Handbook of British Travel Writing*, edited by Barbara Schaff, 95–112. Berlin/Boston: De Gruyter, 2020.

Langbauer, Laurie. *Women and Romance. The Consolations of Gender in the English Novel.* New York: Cornell University Press, 1990.

doi Leask, Nigel. "Eighteenth-Century Travel Writing." In: *The Cambridge History of Travel Writing*, edited by Das, Nandini and Youngs, Tim, 93–107. Cambridge: Cambridge University Press; 2019.

Lidström, Carina. *Berättare på resa. Svenska resenärers reseberättelser 1667–1829*. Stockholm: Carlson, 2015.

Marken, Amy van. "Johannes Meerman. A Dutch traveller to the North and North-East of Europe (1797–1800)." In *Baltic Affairs. Relations between the Netherlands and North-Eastern Europe 1500–1800*, edited by J. Ph. S. Lemmink and J. S. A. M. van Koningsbrugge, 229–48. Nijmegen: INOS, 1990.

doi Meyer, Michael. "Mary Wollstonecraft and the Body of Her Letters, or The Traveler Lost and Found in Scandinavia." In *Traveling Bodies. Interdisciplinary Perspectives on Traveling as an Embodied Practice*, eds. Nicole Maruo-Schröder, Sarah Schäfer-Althaus, Uta Schaffers (New York: Routledge, 2023), 84–89.

doi Radu, Anca-Raluca. "Mary Wollstonecraft, Letters Written in Sweden, Denmark and Norway, 1796." In *Handbook of British Travel Writing*, edited by Barbara Schaff, 267–95. Berlin/Boston: De Gruyter, 2020.

Salih, Sara. "On Judith Butler and Performativity." In Sara Salih, Judith Butler. *Routledge Critical Thinkers*, 55–68. London & New York: Routledge, 2002.

doi Sandrock, Kirsten. "Discourses of Travel Writing." In *Handbook of British Travel Writing*, edited by Barbara Schaff, 31–54. Berlin/Boston: De Gruyter, 2020.

doi Schaff, Barbara, ed. *Handbook of British Travel Writing*. Berlin/Boston: De Gruyter, 2020.

Simonutti, Luisa, "Elsewhere. Women Translators and Travellers in Europe and the Mediterranean Basin in the Age of Enlightenment." In *Gender and Cultural Mediation in the Long Eighteenth Century. Women across Borders*. Mónica Bolufer, Laura Guinnot-Ferri, Carolina Blutrach (Eds.), 193–221. London: Palgrave Macmillan, 2024.

Spies, Marijke. *Bij Noorden om. Olivier Brunel en de doorvaart naar China en Cathay in de zestiende eeuw*. Amsterdam: Amsterdam University Press, 1994.

Wollstonecraft, Mary, William Godwin. *A Short Residence in Sweden and Memories of the Author of 'The Rights of Women'*, edited by Richard Holmes. Penguin, 1987.

doi Zold, Elizabeth. "Motherhood and the Embodied Traveler in Wollstonecraft's Letters Written during a Short Residence in Sweden, Norway, and Denmark." In *Traveling Bodies. Interdisciplinary Perspectives on Traveling as an Embodied Practice*, eds. Nicole Maruo-Schröder, Sarah Schäfer-Althaus, Uta Schaffers (New York: Routledge, 2023), 47–63.

CHAPTER 3

The temporalities of cultural transfer

Robert Louis Stevenson's Pacific travel writing

Kirsten Sandrock
University of Würzburg

This chapter analyses forms and functions of cultural transfer in Robert Louis Stevenson's Pacific travel writings with a particular focus on the temporalities of cultural exchanges. It examines the multiple timeframes of cultural transfer so as to open up a perspective on Stevenson's works as negotiating not only the ambiguities of geographical but also of time-based contact zones, where past, present and future meet. Such an exploration adds a temporal perspective to Mary Louis Pratt's term of the 'contact zone' insofar as it suggests that asymmetrical relationships cannot only arise out of intercultural geographical spaces but can also result from inner-cultural encounters with different periods and times. Ghost imagery and the narration of loss are part of the aesthetic repertoire Stevenson uses in his work to mediate the temporalities of cultural contact zones. Together with other forms of cultural transfer, such as linguistic translation, the construal of the traveller-narrator as intermediary figure and anthropological depictions of Pacific cultures, ghost images and tropes of loss illustrate the complex chronotopical entanglements of cultural encounters that Stevenson negotiates in his Pacific writings. Stevenson's *A Footnote to History* (1892) and *In the South Seas* (1896) offer rich resources for understanding the temporal complexities of cultural transfer because, during his time in the Pacific, Stevenson was deeply occupied with the multiple temporalities of cultures and cultural contact zones.

Keywords: Robert Louis Stevenson, Pacific, cultural transfer, temporality, travel writing, *A Footnote to History*, *In the South Seas*

Travel writers often act as cultural mobilisers. Not always, but frequently in the history of travel writing, authors have adopted an active role in "creating the contact zones of the 'old' world" as well as the 'new' worlds Europeans explored

https://doi.org/10.1075/fillm.20.03san

from the fifteenth century onwards.[1] The term contact zone refers back to Mary Louise Pratt's understanding of the term as a space where different cultures meet, and where they contend with each other in often highly unequal relationships, as it was the case in colonial spaces.[2] In these contact zones, Pratt argues, cultural, linguistic, and social encounters are always connected to a struggle for power, and travel writing as a genre is a particularly powerful source to study the arts of the contact zone.[3] Travel writing is also a powerful source to study the arts of cultural transfer. In colonial times but also beyond that, the transportation, or the facilitation of transportation of cultural knowledges, objects, languages, traditions, and other goods from one part of the world to another is part and parcel of travel writing as a genre.[4] Often, this mobilisation of cultural exchange is conceived of in terms of geographical space: "Boarding a plane, venturing on a ship, climbing onto the back of a wagon, crowding into a coach, mounting on horseback, or simply setting one foot in front of the other and walking."[5] In the majority of cases, spatial movement is conceived of predominantly in such terms of geographical movement. Over the past decades, though, there has been a shift in studies of cultural mobility from affirmative concepts such as "hybridity, network theory, and the complex 'flows' of people, goods, money and information" to the realisation that scholars "need to rethink fundamental assumptions about the fate of culture in an age of global mobility."[6] This article suggests that temporality is one of those aspects to reconsider in order to understand the complexities of cultures and cultural transfer. It proposes that works of travel writing by Robert Louis Stevenson offer a rich resource to study these temporal complexities of the contact zone because they are, like other writings by Stevenson of the time, acutely aware of and concerned with the multiple histories of people, cultures, and natural spaces co-existing at a particular

1. Barbara Korte, "Practices and Purposes," in *Handbook of British Travel Writing*, ed. Barbara Schaff (Berlin: De Gruyter, 2020), 96.

2. Mary Louise Pratt, *Imperial Eyes: Travel Writing and Transculturation* (London: Routledge, 1992).

3. Pratt, *Imperial Eyes.*

4. For an overview of British travel writing, see Barbara Schaff, ed., *Handbook of British Travel Writing* (Berlin: De Gruyter, 2020).

5. Stephen Greenblatt, "A Mobility Studies Manifesto," in *Cultural Mobility: A Manifesto*, ed. Stephen Greenblatt, with Ines G. Županov, Reinhard Meyer-Kalkus, Heike Paul, Pál Nyíri, and Friederike Pannewick (Cambridge: Cambridge University Press, 2010), 250.

6. Stephen Greenblatt, "Cultural Mobility: An Introduction," in *Cultural Mobility: A Manifesto*, ed. Stephen Greenblatt, with Ines G. Županov, Reinhard Meyer-Kalkus, Heike Paul, Pál Nyíri, and Friederike Pannewick (Cambridge: Cambridge University Press, 2010), 1.

time and space.[7] Methodologically, such an exploration adds a temporal perspective to Mary Louis Pratt's term of the 'contact zone' insofar as it suggests that asymmetrical relationships cannot only arise out of intercultural geographical spaces but can also result from inner-cultural encounters with different periods and versions of cultures across time.

Robert Louis Stevenson (1850–1894) was born in Edinburgh, Scotland, in 1850. He was raised and educated in Scotland and England. His father, Thomas, would have liked to see him become a lighthouse engineer. Although he first studied engineering at Edinburgh University, Stevenson eventually decided against becoming an engineer and turned into a "literary star of his generation,"[8] authoring books such as *Treasure Island* (1883), *Kidnapped* (1886) or *Strange Case of Dr Jekyll and Mr Hyde* (1886). From his childhood onwards, Stevenson suffered from ill health and was frequently sick with fevers, coughing, and bronchial problems. He started travelling extensively from the 1870s onwards, partly to flee the Victorian society he grew up in, and partly for medical reasons. A warmer climate was considered to be a possible remedy for his life-long lung problems. His early travels took him to different parts of Scotland and Europe, including the French Riviera. On one of his travels, he met his later wife, American-born author and magazine-writer Fanny Van de Grift Osbourne, whom he married in 1880 in California. After a brief return to Great Britain and then the US again, the couple set out to the South Pacific in 1888. His father, Thomas, had died a year earlier, leaving Robert Louis Stevenson in a deeply melancholic state when embarking on his Pacific travels.[9] This melancholic state may well be one of the reasons why Stevenson's travel writings offer such a multi-layered portrait of the South Pacific with various temporalities of mourning and remembering co-existing in the texts.

This chapter probes into the temporal dimensions of cultural transfer by analysing Stevenson's travel writings, particularly his Pacific works *A Footnote to History* (1892) and *In the South Seas* (1896). In their own time, these works were not perceived as prominent, but they received widespread critical attention over the past decades, i.e., almost a hundred years after they were first written. The rise of travel writing studies and postcolonial studies underlined the uneasy links

7. For another discussion of Stevenson's South Pacific travel writing, on which this one partly expands, see Kirsten Sandrock, "Melancholia in the South Pacific: The Strange Case of Robert Louis Stevenson's Travel Writing," in *The Literature of Melancholia: Early Modern to Postmodern*, eds. Martin Middeke and Christina Wald (Houndmills: Palgrave Macmillan, 2011), Also see, Oliver S. Buckton, *Cruising with Robert Louis Stevenson: Travel, Narrative, and the Colonial Body* (Athens: Ohio UP, 2007), 147–59.

8. Roger Luckhurst, "Introduction," in Robert Louis Stevenson, *The Strange Case of Dr Jekyll and Mr Hyde and Other Tales* (Oxford: OUP, 2006), vii.

9. Sandrock, "Melancholia in the South Pacific," 147–59.

between cultural mobility, cultural transfer, and modernity that are also at the heart of Stevenson's narratives. Adding to existing research on Stevenson's travel writing, this chapter suggests that *A Footnote to History* (1892) and *In the South Seas* (1896) emphasise the entanglements of time and place in the study of cultural exchange and mobility. As most travel writing, they blend genres and function as a narrative site of encounters: not only of cultures and intercultural exchanges but also encounters between different periods and temporal spaces. A focus on the temporalities of cultural transfer provides an additional model to think about one of the grand questions of travel writing and cultural transfer: how to process and communicate the multiple ambiguities of loss and enrichment that are part of cultural contact zones.

A Footnote to History and *In the South Seas* illustrate that cultural contact zones can best be understood as complex chronotopical entanglements, which consist of the active intertwinement of multiple temporal and geographical dimensions. Reflecting critically on the history of European colonialism in the Pacific and Asia, Stevenson's works illustrate that there are multiple Pacifics in one and the same space that have developed out of coexisting and sometimes contradictory historical developments. Through both formal and aesthetic choices, *A Footnote to History* and *In the South Seas* explore these multiple temporalities of the Pacific and introduce readers to the different spaces of cultural encounters. To build up this argument, the chapter begins with a discussion of the forms and functions of cultural transfer in *A Footnote to History* and *In the South Seas*. In a second step, it discusses how temporalities in Stevenson's Pacific writings are poised between past, present and future. The temporal ambiguities also reflect on the ambiguities of cultural transfer in *A Footnote to History* and *In the South Seas*, which view exchanges between Western and Pacific cultures simultaneously as a process of progress and regress, enrichment and loss, newness and demise. Stevenson's travel writings offer a particularly rich source for understanding these multi-temporal spaces and the ways in which travelling in place is also always a matter of travelling in time.

Travel writing as cultural mobiliser

Stevenson lived in the Pacific from the late 1880s onwards and eventually died there in 1894. His published works from this time include *A Footnote to History: Eight Years of Trouble in Samoa* (1892) and *In the South Seas* (1896), the latter of which was originally a series of letters, articles and essays published in the

London *Black and White* and the New York *Sun*.[10] Posthumously, these letters, newspaper articles and essays were published in book form under the title *In the South Sea*. Roslyn Jolly has remarked that the aesthetics of Stevenson's literary output changed around the time he moved to the Pacific. "Even in fiction and travel-writing – the genres which, along with the essay, were most closely identified with his name until 1887 – Stevenson embarked on a radical change of course, shifting from romance to realism, and from the domain of the sentimental traveller to that of the anthropologist."[11] *In the South Seas and A Footnote to History* both reflect this change in narrative form. And yet, certain elements of fiction and the aesthetics of the spiritual continue to shape Stevenson's work and lead to the complex, often ambiguous representation of the spaces the traveller-narrator describes. Material and spiritual worlds meet in Stevenson's texts, which may be partly due to the author's own experiences of the intersections of physical and spiritual realities.

As mentioned earlier, Stevenson had suffered from ill health from childhood onwards. The author's lungs were a particular source of health issues for long periods of his life. After Stevenson had toured central and southern Europe and different parts of the UK and USA, his doctors "suggested that I should try the South Seas."[12] Stevenson journeyed to various Pacific islands, including the Marquees Islands, the Gilbert Islands, and New Caledonia. In 1890, he settled in Samoa, in the village of Vailima, where he died in 1894. At that time, the British Empire had expanded globally and so had other European empires, including those of Spain, France, and Germany. In Samoa and other Pacific islands, the USA was also starting to become active.

Samoa was the scene of conflict, sometimes violent, between three empires, the German (represented by the so-called 'German Firm', owners of large plantations worked by imported labour), the British and the emergent American. Within this context of rival colonial powers three principal leaders of the Samoans, Laupepa, Tamasese and Mataafa, attempted to establish and maintain their own power, though subject to constant interference, particularly from the Germans.[13]

---

10. Neil Rennie, "Introduction and Notes," in Robert Louis Stevenson, *In the South Seas*, ed. Neil Rennie (London: Penguin, 1998), xxii.

11. Roslyn Jolly, *Robert Louis Stevenson in the Pacific: Travel, Empire, and the Author's Profession* (Farnham: Ashgate, 2009), 25.

12. Robert Louis Stevenson, *In the South Seas* [1896], ed. Neil Rennie (London: Penguin, 1998), 5.

13. Graham Tulloch, "A Footnote to History: Stevenson, the Past and the Samoan Present," in *Robert Louis Stevenson and the Great Affair: Movement, Memory, and Modernity*, ed. Richard J. Hill (New York: Routledge, 2017), 148.

The colonial background of the Pacific constantly concerned Stevenson, who was himself part of a European travelling elite but who was, at the same time, deeply concerned about a potential loss of purity that non-European cultures might experience due to overseas travellers, like himself.[14] This ambiguous attitude towards travelling also informed his understanding of cultural transfer and the multiple Pacifics he came to know and write about. To explicate these links, it is useful to first concentrate on the processes of cultural transfer in Stevenson's travel writing and discuss the role the acquisition, processing, and mediation of cultural knowledges played in Stevenson's Pacific writing.

In both *A Footnote to History* and *In the South Seas*, the traveller-narrator construes a literary persona that acts as mobiliser of cultural transfer. *A Footnote to History* partly does so by using the generic "we" in its opening pages. The pronoun suggests that the travel-narrator belongs to the same social and cultural community as the audiences, i.e., an English-speaking, educated, affluent or middle-class readership with interest in the Pacific. The narrator thus functions as a bridge between the British and North American audiences who were the primary readers of Stevenson's books,[15] and the Pacific cultures he writes about. The traveller-narrator emphasises that, although he has been living in the Pacific for some time, he is still part of the West: "To us, with our feudal ideas, Samoa has the first appearance of a land of despotism."[16] The sentence, which imagines a cohesiveness among the English-speaking audience that is ironic in its emphasis on feudalism, works as an entry point into the story of Samoa. It illustrates that it is just as impossible to speak about a collective 'we' in Samoa as it is in Western society, where few people would call themselves feudal in the late nineteenth century. Similarly, Samoa emerges not as a land of despotism, but one of complex cultural traditions: "An elaborate courtliness marks the race alone among Polynesians; terms of ceremony fly thick as oaths on board a ship; commoners my-lord each other when they meet – and urchins as they play marbles."[17] Stevenson's narrative uses an insider-outsider perspective to create a bridge between the traveller-narrator, the people he writes about and the audience he writes for. If cultural mobilisers are "agents, go-betweens, translators, or intermediaries" who "facilitate contact" between cultures,[18] then Stevenson's narrator functions as such an intermediary figure who uses the first-person plural to

14. Sandrock, "Melancholia in the South Pacific," 147–59.

15. Tulloch, "A Footnote," 156.

16. Robert Louis Stevenson, *A Footnote to History: Eight Years of Trouble in Samoa* (Dodo Press, 2007), 1.

17. Stevenson, *A Footnote*, 1.

18. Greenblatt, "A Mobility Studies Manifesto," 251.

mark his place as an insider of both Western and Pacific cultures. The narrator moves between worlds, and readers are invited to move with the narrator and explore. It was Stevenson's pronounced aim to "serve Samoa and Samoans" by writing *A Footnote to History*, which was meant "to attract readers and then to influence them and, through them, to influence imperial policy."[19] Travel writing here has an overt political objective, and so do the processes of cultural transfer in Stevenson's travel narratives.

In the South Seas makes an even more explicit case for the intermediary function of the traveller-narrator. It is a collection of articles that cover Stevenson's travels in the Marquesas, the Paumotus (today called Tuamotus), the Gilbert Islands and Abemama, which is one of the Gilbert groups in Kiribati. The processing and mediation of the cultural knowledges gained on these Pacific travels is, in the words of the traveller-narrator, one of the main aims of writing *In the South Seas*. The second paragraph contains the following passage:

> the task before me is to communicate to fireside travellers some sense of its [the Pacific's; K. S.] seduction, and to describe the life, at sea and ashore, of many hundred thousand persons, some of our own blood and language, all our contemporaries, and yet as remote in thought and habit as Rob Roy or Barbarossa, the Apostles and the Caesars.[20]

The sentence identifies writing as a primary tool of cultural transfer. It exemplifies that "literary texts [are] also used in the mobilisation of a people,"[21] and that by means of writing, authors can try to overcome temporal and geographical distances. The passage singles out the target audience of the book as "fireside travellers," referring to Stevenson's mostly European and North American audiences, what Graham Tulloch refers to as the author's "northern hemisphere readers."[22] For these audiences, the traveller-narrator suggests, the world of the Pacific is "as remote in thought and habit as Rob Roy or Barbarossa, the Apostles and the Caesars." At the same time, some of the people in the Pacific are described as being "of our own blood and language," which creates ethno-lingual ties as part of the cultural networks in the nineteenth-century Pacific. The distance between the Pacific and the West is expressed in terms of temporal remoteness, and the

19. Tulloch, "A Footnote," 150.

20. Stevenson, *In the South Seas*, 5–6.

21. Petra Broomans, "The Importance of Literature and Cultural Transfer – Redefining Minority and Migrant Cultures," in *Battles and Borders: Perspectives on Cultural Transmission and Literature in Minor Language Areas*, eds. Petra Broomans, Goffe Jensma, Ester Jiresch, Janke Klok, and Roald van Elswijk (Groningen: Barkhuis, 2015), 14.

22. Tulloch, "A Footnote," 156.

narrator becomes a traveller in time as much as in place. Seventeenth-century Scotland, when Robert Roy MacGregor – "Rob Roy" – lived, the medieval Germany of Frederick I, Holy Roman Emperor, also known as Frederick Barbarossa, the biblical age of the Apostles or the classical Rome of Caesar: all of these are said to be as remote as Pacific cultures are to contemporary readers. It is the narrator's self-proclaimed task to "communicate" to readers what this part of the earth is like. From the opening paragraphs onwards, *In the South Seas* opens up such multiple spatial paradigms that depict the processes of cultural transfer as both temporal and topographical exchanges. Stevenson's travel works indicate that the processing and mediation of cultural experiences is never purely a transfer in topographical terms. It also involves temporal travelling because cultures inhabit different time-zones, and oftentimes several at once.

Other practices of cultural transfer in *A Footnote to History* (1892) and *In the South Seas* include practices of linguistic translation, the interpretation of religious practices and cultural traditions, the collection and sharing of historical information, close descriptions of the landscapes and its people as well as the frequent comparisons between Pacific and British cultures, especially Scottish Highland culture. For Tulloch, there are "four main perspectives" that interact in the works: "the historical, the geographical, the literary, and the psychological."[23] All of these work towards the processes of cultural transfer. Both *In the South Seas* and *A Footnote to History* contain passages where Stevenson translates phrases, sayings, or conversations from native Pacific languages into English. *A Footnote to History* mostly introduces single terms and explains what these terms mean to Samoan people. Oftentimes, the traveller-narrator introduces terms into English for which no equivalent English word exists. The text then explains these linguistic gaps by offering details of the cultural context. The following passage introduces readers to the concept of *malanga* and illustrates the anthropological and linguistic forms of cultural transfer:

> But the special delight of the Samoan is the *malanga*. When people form a party and go from village to village, junketing and gossiping, they are said to go on a *malanga*. Their songs have announced their approach ere they arrive; the guest-house is prepared for their reception; the virgins in the village attend to prepare the kava bowl and entertain them with the dance; time flies in the enjoyment of every pleasure which an islander conceives; and when the *malanga* sets forth, the same welcome and the same joys expect them beyond the next cape, where the nearest village nestles in its grove of palms. To the visitors it is all golden; for the hosts, it has another side. In one or two words of the language the fact peeps slyly out. The same word (*afemoeina*) expresses "a long call" and "to come as a

23. Tulloch, "A Footnote," 152.

> calamity"; the same word (*lesolosolou*) signifies "to have an intermission of pain" and "to have no cessation, as in the arrival of visitors"; and *soua*, used of epidemics, bears the sense of being overcome as with "fire, flood, or visitors." But the gem of the dictionary is the verb *alovao*, which illustrates its pages like a humorous woodcut. It is used in the sense of "to avoid visitors," but it means literally "hide in the wood." So, by the sure hand of popular speech, we have the picture of the house deserted, the *malanga* disappointed, and the host that should have been quaking in the bush.[24]

Linguistic translation here goes hand in hand with cultural translation: first, words are literally translated as part of linguistic transfer while, second, this linguistic transfer opens up the space for cultural transfer, where concepts and cultural practices are translated and introduced. In this process, readers get a sense of the complexities of both indigenous language and cultural practices. To introduce *malanga* to readers, to explain the significance of going "from village to village, junketing and gossiping" and then to comment on the ambiguity of the custom for different participant in the ritual may not introduce an explicit temporal dimension to the processes of cultural transfer, but it illustrates that, as in every culture, several realities coexist in one and the same place at the same time. For visitors, *malanga* is a joyful practice, an "all golden" activity in which spatial movement is linked to social exchange and communal conviviality. For the hosts, it is linked to stasis and forced hospitality, starting and ending with a sense of "calamity" that perhaps not only people in the Pacific can relate to but also Western readers. In passages like this, *A Footnote to History* interprets what Clifford Geertz called the "semiotic" processes of culture and its "webs of significance" that people themselves have "spun."[25] Travel writing, for Stevenson, is not merely about subjective experiences. It is about the interpretation and translation of cultures so that readers will be able to decipher at least some of the words, signs, and traditions that are part of the meaning-making processes of Pacific cultures.

A further example helps to illustrate the facets of Stevenson's Pacific writings and the processes of cultural transfer they participate in. *In the South Seas* translates not only words and phrases from Samoan into English but sometimes also passages from Pacific poetry or song. Chapter 3 sets out to describe "the beauties of Anaho" and uses the words from "the Hawaiian poet" – who, according to Neil Rennie, has never been identified[26] – to transmit the beauty of Anoha to readers in local literary aesthetics:

24. Stevenson, *A Footnote to History*, 5–6.

25. Clifford Geertz, "Thick Description: Toward an Interpretive Theory of Culture," in Clifford Geertz, *The Interpretation of Cultures: Selected Essays* (New York: Basic Books, 1973), 5.

26. Rennie, Introduction and Notes, 260.

> *Ua maomao ka lani, ua kahaea luna,*
> *Ua pipi ka maka o ka hoku*
> (The heavens were fair, they stretched above,
> Many were the eyes of the stars.)[27]

The personification of the sky and the stars conveys a sense of beauty that links Pacific nature to expansiveness, brightness, and pleasantness. It also conveys a sense of timelessness that is crucial to Stevenson's depiction of the Pacific. The sky and the stars are perhaps not eternal, but they come close to being immortal entities compared to human beings. They are also, to a certain extent, transregional and symbolise the connectedness of the world when it comes to larger spatial frameworks.

Not coincidentally, this reflection on the Pacific sky and the stars leads the traveller-narrator to reflect a few lines later on the Scottish Highlands, a comparison which frequently enters Stevenson's text for various purposes. Ann Colley writes about Stevenson's time in the Pacific that "Scotland and Empire were both available through these reminders of home, but, paradoxically, Scotland was even more valuable as a point of reference that helped him understand his new surroundings. [...] Scotland offered parallels by which, for instance, he could better understand the structure of island society (through its clan system) and its political dilemmas."[28] In the passage cited above, the Hawaiian song induces the traveller-narrator to imagine he "had slipped ten thousand miles away and was anchored in a Highland loch."[29] The link between the localities appears natural to the traveller-narrator, who mentally likens the Pacific island to the Scottish Highlands. There are numerous instances where such comparisons between Pacific and Scottish culture enter Stevenson's works and open up a connection between the physical and mental environments of Europe and the Pacific. Significantly, many of these comparisons do not reflect on the nineteenth-century Scotland Stevenson had grown up in, but on a "Scottish past" that signifies both a distant place and a distant time for Stevenson.[30] References to the Highlands operate at the interfaces of time and place. They tap into an aesthetics that forges a connection between geographical and temporal remoteness, thus evincing concern with the multiple Pacifics the traveller-narrator experiences while living there.

---

27. Stevenson, *In the South Seas*, 17.

28. Ann C. Colley, *Robert Louis Stevenson and the Colonial Imagination* (Hampshire: Ashgate, 2004), 5.

29. Stevenson, *In the South Seas*, 18.

30. Colley, *Robert Louis Stevenson*, 5.

Cover with the courtesy of Wordsworth Classics

The temporalities of cultural transfer

Long before his Pacific journeys, travel writing had been one of the genres Stevenson flourished in. From his first travels in the UK and Europe, he published works that relate his travel experiences to his readers while also introducing an English-speaking audience to different sides of Scotland, Belgium, France, New York, California, and the Pacific. His better-known travelogues from earlier travels include *An Inland Voyage* (1878) about Belgium and northern France, *Edinburgh: Picturesque Notes* (1878), *Travels with a Donkey in the Cévennes* (1879), and *Across the Plains* (1892), which relates his journey from New York to California by train. As Richard J. Hill notes, "Stevenson's literary works reveal an author interested in the effects of movement, the narrative possibilities, and pitfalls, of characters moving rapidly from place to place."[31] In addition to movement from place to place, Stevenson's travel narratives are also interested in temporal movement. This emerges, in addition to the examples above, in the tropes of loss and the ghost imagery that pervade his works.

To start with the sense of loss, Stevenson's Pacific writings have been identified as contributing to a period of British travel writing in the late nineteenth and early twentieth centuries that was centrally concerned with loss.[32] This has several reasons. First, some authors questioned the processes of industrialisation and capitalism that had shaped British and European society for the past centuries, and they saw in some of the other cultures and countries they travelled a sense of purity or originality that they deemed lost in Europe. Second, and partly related to this, the role of the self in relation to its surroundings was profoundly questioned in the nineteenth and twentieth centuries. As in other arts and literary genres of the time, there was a loss of individual self-assurance and also a loss of national self-assurance, which was partly expressed through "[a] programmatic writing of the fragile self in the light of experiencing the often precarious other."[33] Several scholars have conceptualised this sense of loss and self-disintegration in Stevenson's travel writing in psychoanalytic terms, drawing, amongst others, on theories of mourning, melancholy, and masochism.[34] One aspect that plays into these readings is the temporality of loss in Stevenson's travel texts, which is superimposed in the tropes of death, demise, and melancholia.

In the South Seas is a prime specimen of negotiating the self-other relationship in light of loss. This ranges from the loss of personal health to the loss of young age and the loss of future generations.[35] *A Footnote to History* and *In the South*

31. Richard J. Hill, "Introduction: Robert Louis Stevenson and the Great Affair: Movement, Memory and Modernity," in *Robert Louis Stevenson and the Great Affair: Movement, Memory, and Modernity*, ed. Richard J. Hill (New York: Routledge, 2017), 2.

32. Barbara Schaff, "Periods of Travel Writing," in *Handbook of British Travel Writing*, ed. Barbara Schaff (Berlin: De Gruyter, 2020), 22.

33. Schaff, "Periods of Travel Writing," 22.

34. E.g., John Kucich, "Melancholy Magic: Masochism, Stevenson, Anti-Imperialism," *Nineteenth-Century Literature* 56, no. 3 (2001): 364–400; Sandrock, "Melancholia in the South Pacific," 147–59.

Seas both take issue with the demise that has set in among Pacific cultures after the period of European contact. They particularly discuss changes in Pacific cultures due to European colonialism and conceptualise these changes in terms of multiple temporalities. *A Footnote to History: Eight Years of Trouble in Samoa*, for one, relates in detail the history of colonisation in Samoa between 1882 and 1892. Its very title and subtitle emphasise the significance of temporality, with both "*History*" and "*Eight Years*" being time-based signifiers. In the years depicted in the book, German, US American, and British forces tried to get control over the island and to use it as a political and commercial trading post. The text opens with a sentence that emphasises the timeliness of the story Stevenson sets out to write. In so doing, it links the concept of history-writing to the presence while also suggesting that colonisation is an ongoing process: "The story I have to tell is still going on as I write; the characters are alive and active; it is a piece of contemporary history in the most exact sense. And yet, for all its actuality and the part played in it by mails and telegraphs and iron war-ships, the ideas and the manners of the native actors date back before the Roman Empire."[36] Past and present intermingle in this excerpt, which links history to story-telling more generally, and the contemporary Pacific to the ancient Roman Empire in specific. The references to relatively recent technological innovations – the telegraph, iron war ships and overseas mail services – signal the nineteenth-century colonial history of the Pacific, which seems to clash with the more traditional lifestyles of local cultures prior to colonisation.

Colonial encounters actively changed Pacific island cultures and led to the coexistence of multiple timeframes on the islands. Chapter II of *A Footnote to History* is titled "The Elements of Discord: Foreign," and it discusses how Western conceptions of progress and profitability have lastingly changed Samoan ways of life, whether locals wanted it or not:

> The huge majority of Samoans, like other God-fearing folk in other countries, are perfectly content with their own manners. And upon one condition, it is plain they might enjoy themselves far beyond the average of man. [...] But the condition – that they should be let alone – is now no longer possible. More than a hundred years ago, and following closely on the heels of Cook, an irregular invasion of adventurers began to swarm about the isles of the Pacific. The seven sleepers of Polynesia stand, still but half aroused, in the midst of the century of competition. And the island races, comparable to a shopful of crockery launched upon the stream of time, now fall to make their desperate voyage among pots of brass and adamant.[37]

35. Sandrock, "Melancholia in the South Pacific," 147–59.

36. Stevenson, *A Footnote to History*, 1.

The effects of the cultural contact zones are not only the loss of autonomy but also the impossibility of turning back time: the "condition – that they should be let alone – is now no longer possible." Together with European contact and the transfer of knowledges between Pacific and European cultures, Samoans lost a level of sovereignty. They are stuck on the global "stream of time" that Western cultures have increasingly imposed on the world through expansion schemes and which can be read as an allegory of linear concepts of time. This is the dilemma of the contact zones for Stevenson. One may criticise colonial power dynamics and recognise one's own place in this, as Stevenson later does. Yet, it is impossible to take back the changes brought onto Pacific cultures. The indigenous population of Samoa may have been able to drive out German colonisers, who had been settling in the Samoan Islands and the Central Pacific since the mid-nineteenth century. They could not, though, escape the increasing normativity of linear time that Western modernity brought on to cultures around the world by means of global expansion.

One tool to resist this development, for Stevenson's narrator, is the invocation of ghosts as both spiritual and religious entities that are, by definition, unbound by place and time. In his fiction, the author sometimes used "romantic and Gothic aesthetics," thereby "displacing the concept of time as static, and disrupting the notion of a unified human identity."[38] In travel writing, Gothic elements and particularly the element of the ghost feature similarly strongly, and they emphasise the multiple timeframes of Stevenson's Pacific writings. Ghosts may not be the most typical emblems of cultural transfer. Yet, in Stevenson's work, they take up the role of intermediary figures that move between worlds, above all temporal worlds. In the opening lines of *In the South Seas*, the ghost is an ironic, yet complex symbol of the traveller. It stands for the traveller-narrator himself, but also for the spatial entanglements of dwelling in multiple temporalities at once:

> For nearly ten years my health had been declining; and for some while before I set forth upon my voyage, I believed I was come to the afterpiece of life, and had only the nurse and undertaker to expect. It was suggested that I should try the South Seas; and I was not unwilling to visit like a ghost, and be carried like a bale, among scenes that had attracted me in youth and health.[39]

37. Stevenson, *A Footnote to History*, 9.

38. Bridget Mellifont, "Temporality and Text: Shapes of Time in R. L. Stevenson's 'The Beach of Falesá' and 'Markheim'" in *Robert Louis Stevenson and the Great Affair: Movement, Memory, and Modernity*, ed. Richard J. Hill (New York: Routledge, 2017), 73.

39. Stevenson, *In the South Seas*, 5.

In terms of geographical space, the passage relates a fairly straightforward movement. The traveller-narrator was advised to journey to the Pacific, which he did. In terms of temporal movement, the passage is much more complex. It moves from the recent past ("For nearly ten years") to the present ("my voyage") to the imagined future ("the afterpiece of life") and back to the distant past ("in youth and health"). These multiple temporal worlds are connected through the image of the ghost, which personifies, by means of simile, the narrator himself. The image reflects, on the one hand, the prominent "melancholy awareness" of travel writers in the late nineteenth and early twentieth century, many of whom witnessed and critically discussed how "cultures were vanishing, disintegrating and becoming irrevocably lost."[40] On the other hand, the ghost image invites a sustained look at the multiple timeframes ingrained in the processes of melancholy and mourning and also in the processes of cultural transfer in Stevenson's writings. Not only are the objects that are deemed to be lost in the past carried into the present by means of mourning. Travel writing, as one process of cultural transfer, also tries to preserve the lost objects for the future by putting them down on paper.

Stevenson was well aware of the ambiguities of trying to affix cultural meaning in writing at the same time as resisting the idea that any kind of stable identity of cultures may be possible.[41] After all, writing is frequently a means to fasten the otherwise unstable processes of the contact zones in a more or less stable material form. Printed books, handwritten or typed letters, diaries or other forms of travel writing usually communicate the real or imagined journeys of people, often in aestheticised form. The literary work as a physical object is partly fixed in space and time, which seems to contradict the idea of a genre for which movement and cultural exchanges are *sine qua non*. As usual when it comes to ambiguities, Stevenson was well aware of these paradoxes and made them part of the aesthetic construction of his work. In earlier works, above all in *Strange Case of Jekyll and Hyde* (1886), Stevenson elaborated on the ambiguities of Victorian culture. In his Pacific writings, he presents readers with the ambiguities of cultural encounters and of processes of cultural transfer. By embedding different chronologies in his texts, Stevenson's travel narratives exemplify the inherently ambiguous nature of literary works. They resist fixed perspectives, which makes the processes of cultural transfer in the texts partly unpredictable, yet all the more productive. Ghost imagery is part of this negotiation and a powerfully anachronistic symbol resisting the ideological and discursive paradigms of Western modernity while likewise reminding readers of the impossibilities of getting to know a single or fixed Pacific past.

40. Schaff, "Periods of Travel Writing," 22.

41. Hill, "Introduction," 2.

Ghosts are, of course, part of the aesthetic repertoire of literature. They embody the ongoing presence of the past in the present and in the future, especially of those aspects that may have been repressed in the past and which are therefore haunting the present and future all the more powerfully. In relation to Stevenson's travel writing, they aesthetically frame the transhistoric quality of moving in space and of mediating this experience by means of literature. There are several passages in *In the South Seas* where the traveller-narrator dwells on the significance of ghosts in Pacific cultures.[42] Believing in ghosts is common, according to the text, in Pacific cultures. As an outsider, and as an author interested in the past and present, Stevenson interprets the spiritual world of the Pacific in a way that makes ghosts appear as part of an everyday spirituality. He also portrays it as another prime specimen of how different worldviews and temporalities mix in the Pacific: "all men believe in ghosts, all men combine with their recent Christianity fear of and a lingering faith in the old island deities."[43] Following the patterns of Stevenson's earlier works, which one critic describes as "seek[ing] to amplify rather than resolve duality,"[44] Stevenson's travel writings dwell on the coexistence of different belief systems and entire spiritual epistemologies in the nineteenth-century Pacific. Different chronologies coexist in Stevenson's work, and oftentimes these different chronologies are linked to the contact zones of European and Pacific cultures. This can be seen, for instance, in Stevenson's depiction of the opium trade and its long-term effects in the Marquesas, a former opium plantation hub.

Stevenson writes about the history of opium trade and the global dynamics this trade has created by the late nineteenth century. Even though Pacific islands were no longer the primary producer of opium and, by the 1890s, opium "plantations are practically deserted and the Chinese gone,"[45] *In the South Seas* notes the enduring effects global opium trade had on the islands and its people: "in the meanwhile the natives had learned the vice, the patent brings in a round sum, and the needy Government of Papeete shut their eyes and open their pockets."[46] Not unlike a ghost, opium production and consumption continue to disturb Pacific cultures and peoples. Even though travellers in the late-nineteenth century are not directly responsible for the opium trade of earlier times, *In the South Seas* emphasises that everyone in the UK is still indirectly involved in the lasting legacy of the

42. E.g., Stevenson, *In the South Seas*, 27–29; 140–42; 149–51.
43. Stevenson, *In the South Seas*, 142.
44. Luckhurst, "Introduction," xxii.
45. Stevenson, *In the South Seas*, 54.
46. Stevenson, *In the South Seas*, 54.

opium trade and the way it transformed Pacific life. The temporalities of past and present mix in the Pacific's relationship to opium:

> Those that live in glass houses should not throw stones; as a subject of the British crown, I am an unwilling shareholder in the largest opium business under heaven. But the British case is highly complicated; it implies the livelihood of millions; and must be reformed, when it can be reformed at all, with prudence.[47]

The phrase "an unwilling shareholder" underlines the continuity of the opium trade and the ongoing effects it has on the Marquesas. The processes of cultural transfer are hardly ever neutral. *In the South Seas* illustrates that they lead to split temporalities, both for those who visit these parts of the world and for those who live there. "Stevenson reminds us that when one starts to look closely at the life of someone, not himself a coloniser, but living in a territory that is in the process of being colonised by his own country, one is bound to have complicated responses that disqualify the clichés of hindsight and that make the question seem irrelevant and naïve."[48] For Stevenson's depiction of the Pacific, and for his role in the processes of cultural transfer, the emphasis on heterogenous temporality is a means to reflect on the many ambivalences that come with cultural transfer. Stevenson's travelogues convey the temporal divisions produced by colonialism, and they medialise the multiple time-schemes that co-exist in the Pacific partly due to earlier contact zones.

Conclusion

Stevenson's critique of Western colonialism in his Pacific travel writing contributes to an understanding of modernity as having imposed teleological conceptions of time on the West and cultures around the world. As Fredric Jameson argues, the modernist preoccupation with time in Western cultures was a result of a deeply fragmented society, in which numerous people "lived in two distinct worlds simultaneously; born in those agricultural villages we still sometimes characterise as medieval or premodern, they developed their vocations in the new urban agglomerations with their radically distinct and 'modern' spaces and temporalities."[49] Born and raised in Edinburgh, Stevenson had not grown up in a premodern surrounding. Yet, in his final years of life he deliberately moved to a place where he witnessed the multiplicities of time Jameson considers to be character-

47. Stevenson, *In the South Seas*, 54–5.
48. Colley, *Robert Louis Stevenson*, 137.
49. Fredric Jameson, "The End of Temporality," *Critical Inquiry* 29, no. 4 (2003): 699.

istic of the late nineteenth and early twentieth century, and which Jameson also links to the history of European expansion around the time of the Berlin conference in 1885,[50] thus around the time Stevenson wrote about the Pacific. For Stevenson, Samoa and other Pacific cultures are a lens through which to negotiate the clash of different temporal contact zones that also existed in the UK at that time, but which were all the more forcefully felt, at least for the traveller-narrator, abroad. Stevenson's Pacific narratives reflect on larger changes in spatial and temporal epistemologies in the late nineteenth century, when certain forces were trying to impose singular time schemes around the world whereas other forces were increasingly recognising that such a singular timeframe is impossible. Stevenson did not unquestionably believe in a progressive version of history, one that endorses linear conceptions of time.[51] Yet, he saw both in the UK and abroad that alternative conceptions of time were difficult to uphold in the light of increasingly globalised structures. His travel writing reflects on this ambiguity by linking travelling not only to geographical movement but also to temporal movement. They also engage readers in this process and illustrate how cultural transfer is always both a topographical as well as a temporal process.

References

Broomans, Petra. "The Importance of Literature and Cultural Transfer – Redefining Minority and Migrant Cultures." In *Battles and Borders: Perspectives on Cultural Transmission and Literature in Minor Language Areas*, edited by Petra Broomans, Goffe Jensma, Ester Jiresch, Janke Klok, and Roald van Elswijk, 9–38. Groningen: Barkhuis, 2015.

Buckton, Oliver S. *Cruising with Robert Louis Stevenson: Travel, Narrative, and the Colonial Body*. Athens: Ohio University Press, 2007.

Colley, Ann C. *Robert Louis Stevenson and the Colonial Imagination*. Hampshire: Ashgate, 2004.

Geertz, Clifford. "Thick Description: Toward an Interpretive Theory of Culture." In Clifford Geertz, *The Interpretation of Cultures: Selected Essays*, 3–30. New York: Basic Books, 1973.

Greenblatt, Stephen. "Cultural Mobility: An Introduction." In *Cultural Mobility: A Manifesto*, edited by Stephen Greenblatt, with Ines G. Županov, Reinhard Meyer-Kalkus, Heike Paul, Pál Nyíri, and Friederike Pannewick, 1–23. Cambridge: Cambridge University Press, 2010.

50. Jameson, "The End," 699.

51. Jolly, *Robert Louis Stevenson*, 75.

Greenblatt, Stephen. "A Mobility Studies Manifesto." In *Cultural Mobility: A Manifesto*, edited by Stephen Greenblatt, with Ines G. Županov, Reinhard Meyer-Kalkus, Heike Paul, Pál Nyíri, and Friederike Pannewick, 250–3. Cambridge: Cambridge University Press, 2010.

doi Hill, Richard J. "Introduction: Robert Louis Stevenson and the Great Affair: Movement, Memory and Modernity." In *Robert Louis Stevenson and the Great Affair: Movement, Memory, and Modernity*, edited by Richard J. Hill, 1–10. New York: Routledge, 2017.

doi Jameson, Fredric. "The End of Temporality." *Critical Inquiry* 29, no. 4 (2003): 695–718.

Jolly, Roslyn. *Robert Louis Stevenson in the Pacific: Travel, Empire, and the Author's Profession.* Farnham: Ashgate, 2009.

doi Korte, Barbara. "Practices and Purposes." In *Handbook of British Travel Writing*, edited by Barbara Schaff, 95–112. Berlin: De Gruyter, 2020.

doi Kucich, John. "Melancholy Magic: Masochism, Stevenson, Anti-Imperialism." *Nineteenth-Century Literature* 56, no. 3 (2001): 364–400.

Luckhurst, Roger. "Introduction." In Robert Louis Stevenson, *Strange Case of Dr Jekyll and Mr Hyde and Other Tales*, edited by Roger Luckhurst, vii–xxxii. Oxford: Oxford University Press, 2006.

doi Mellifont, Bridget. "Temporality and Text: Shapes of Time in R. L. Stevenson's 'The Beach of Falesá' and 'Markheim.'" In *Robert Louis Stevenson and the Great Affair: Movement, Memory, and Modernity*, edited by Richard J. Hill, 73–88. New York: Routledge, 2017.

Rennie, Neil. "Introduction and Notes." In Robert Louis Stevenson, *In the South Seas* [1896], edited by Neil Rennie, viii–xxxv and 258–82. London: Penguin, 1998.

doi Sandrock, Kirsten. "Melancholia in the South Pacific: The Strange Case of Robert Louis Stevenson's Travel Writing." In *The Literature of Melancholia: Early Modern to Postmodern*, edited by Martin Middeke and Christina Wald, 147–59. Houndmills: Palgrave Macmillan, 2011.

doi Schaff, Barbara, ed. *Handbook of British Travel Writing*. Berlin: De Gruyter, 2020.

doi Schaff, Barbara. "Periods of Travel Writing." In *Handbook of British Travel Writing*, edited by Barbara Schaff, 11–30. Berlin: De Gruyter, 2020.

Stevenson, Robert Louis. *A Footnote to History: Eight Years of Trouble in Samoa* [1892]. Dodo Press, 2007.

Stevenson, Robert Louis. *In the South Seas* [1896], edited by Neil Rennie. London: Penguin, 1998.

doi Tulloch, Graham. "A Footnote to History: Stevenson, the Past and the Samoan Present." In *Robert Louis Stevenson and the Great Affair: Movement, Memory, and Modernity*, edited by Richard J. Hill, 148–61. New York: Routledge, 2017.

CHAPTER 4

Postcolonial images, ambivalence and weak border zones

"Us" and "the others" in the account of an early twentieth-century Swedish traveller

Eduardo Gallegos Krause
Universidad de La Frontera

This chapter deals with the largely unknown and slightly studied travel account of Otto Nordenskjöld in the Argentinian-Chilean Patagonia as part of the preparation to his well-known and documented travel to Antarctica. By using several key terms of postcolonial theory, such as ambivalence, border, identity, binaries and in-between, an analysis of the way in which Nordenskjöld's portrays the image of the Indigenous people and how civilisation treats them (from Nordenskjöld's perspective) is proposed. Thus, it is argued that, in the complex relation between travel account and images, an ambiguous representation of European values and the Indigenous people arises, which destabilises the typical binaries of the colonial discourse. The contribution demonstrates that, as travellers transition between two realms, they may find it challenging to shed their European viewpoint, even in situations where colonial brutality is openly denounced.

Keywords: postcolonial discourse, borders, West-East dichotomy, Indigenous peoples, iconic representations

Nils Otto Gustaf Nordenskjöld (1869–1928), known as Otto Nordenskjöld, was a Swedish geologist, geographer and explorer. He is primarily remembered for leading the Swedish Antarctic Expedition (1901–1904), although he also served as a professor of geography and ethnography at the University of Gothenburg.[1]

1. To avoid confusion, it is worth noting that his cousin, Erland Nordenskiöld, was also a renowned explorer in South America. While Otto was dedicated to the recognition of Chilean-Argentine Patagonia and Antarctica, Erland is remembered for his work in the Bolivian Chaco, and with indigenous people from Peru, Colombia and other areas. See, in this regard, Alvarsson, Jan-Åke and Agüero, Oscar (Ed.), *Erland Nordenskjöld. Investigador y amigo del indígena* (Abya Yala: Quito, 1997).

https://doi.org/10.1075/fillm.20.04gal

These biographical references are important because knowledge of Otto Nordenskjöld, especially in his capacity as geographer, was used not only in the context of the expedition to Antarctica but also in the matter of the rectification of the border between Chile and Argentina,[2] an important issue regarding the postcolonial dimension of this expedition manifested in the interests of Europeans for an international order in terms of occidental values and economics interests.

In this sense, Nordenskjöld's remembrance often highlights his trip to Antarctica, while ignoring the mediator role accomplished by the Swedish in the Chilean-Argentine Patagonia (1894–1897). This mediator role was possible, at least, on three different levels: first, as an arbitrator in the frontier dispute between Chile and Argentina, although not in a formal role, but as a validated scientist. Second, as a representative of modern and civilised manners (in the perspective of his own time) in a territory still inhabited by Indigenous people who, according to the paradigm of that time, kept their traditional manners. The third form of mediation, related to the other two already mentioned, is the way in which Nordenskjöld's journey was communicated by different journals using a complex hybridity between language descriptions and photographs or images. In this regard, the socio-semiotic approach for analysing the image, as will be shown, appears as a useful model to embrace the symbolic value related to the border zone.

Regarding the symbolic dimension of borders, the first two mediations are related to the two meanings associated with the concept of 'border' or 'border zone': on one hand a border signifies a frontier between two states, and on the other, a border zone establishes an edge, a divisionary line which propose a margin and a centre.[3] These two are directly connected with the two first mediations described in the paragraph above. In both of them, the territory, either as state, or a liminal space, or both at the same time, is considered with an inherent symbolic value; as a space full of meaning and related to the people that inhabit it. Evidently, the first mediation, or border as a frontier between two states, deals with a more literal sense and has several implications regarding the expansion of capital which relates to the complicity of travel writing with imperialism, or postcolonialism, as I prefer to call it here, as it has been demonstrated by several scholars.[4]

2. See Maurice Zimmermann, "Amérique du Sud. Rectifications de frontières," *Annales de Géographie*, 17 (1895): 518–9; A. Fournier, "Expédition O. Nordenskjöld à la Terre de Feu," *Annales de Géographie*, 21 (1896): 341–2.

3. T. Youngs, "Border," in *Keywords for travel writing studies*, eds. Ch. Forsdick, Z. Kinsley, K. Walchester (London: Anthem, 2019), 25–27.

4. E. Said, *Orientalism* (London: Penguin, 2003); M.L. Pratt. *Imperial Eyes. Travel writing and transculturation.* (New York: Routledge, 2008); D. Spurr. *The Rhetoric of Empire. Colonial Discourse in Journalism, Travel Writing, and Imperial Administration.* (Duke University Press: Durham, 1993).

The second mediation, or border as a distinction between centre and margin, deals with the definition of territories and populations and the relation to one another: "Travel literature is frequently populated with the representation of borders between the known and the unknown, the civilised and the savage, the domestic and the wild, the land and the sea, the cultivated and the desert."[5] Nevertheless, this definition can be contrasted with the travel experience, as it appears in its linguistic, verbal and iconic image dimension, in order to challenge the idea about borders as strong zones. A strong zone, as suggested in the previous quotation, gives the impression of a total demarcated distinction between, for instance, the civilised and the savage. I refer here to the idea of the noble savage as deployed since the eighteenth century in clear relation with the idea of the savage which worked fruitfully for the hierarchised relationship between Europe and its others.[6] In this sense, the "good savage" works here as an improvement or something not as bad as the savage itself. However, it has been shown that this construction "(...) produces an ostensibly positive oversimplification of the 'savage' figure, rendering it in this particular form as an idealised rather than a debased stereotype."[7]

Image 1. Nordenskjöld, "La terre de feu", 15

5. Youngs, "Border," 25.

6. T. Todorov, *On Human Diversity: Nationalism, Racism, and Exoticism in French Thought.* (Harvard University Press: Massachusetts, 1998).

7. B. Ashcroft, G. Griffiths and H. Tiffin, *Post-colonial studies. The key concepts (Routledge: London, 2007)*, 192–3.

In contrast with this dichotomous idea, I argue that the travel account of Otto Nordenskjöld leads to a problematic description of the civilisational process in which the differences between the civilised and the 'savage' are questioned. Thus, the border zone appears as a weak delimitation with no fixed boundary or distinction between the civilised or the so-called barbarian peoples. This will be considered, particularly, in the relation between image and story that travel writing proposes.

All this is possible because of the travel's cultural mobility and the possibility of contact zones that the travel implies are understood as an encounter between two asymmetric cultures with an interactive dimension. In the words of Mary Louise Pratt, contact zones are: "(...) social spaces where disparate cultures meet, clash, and grapple with each other, often in highly asymmetrical relations of domination and subordination (...)".[8] But this definition emphasises a binary relation in which the asymmetry prevails, instead of that, I propose that the asymmetrical relations can be contested. Thus, and as I will demonstrate, it is in the link between image and linguistic description that the binaries can be dissolved in a more ambiguous manner. With regard to the photographs in question, there are two images in Nordenskjöld's travel account that will be analysed here. The first one (Image 1) shows all the European expeditioners who went with Nordenskjöld in his journey. The second image (Image 2) shows a group of Indigenous people from the Rio Grande mission. This, I argue, represents a part of a civilisation effort, characterised by the work of religious missionaries. Yet, at the same time, this effort is confronted with some contradictions regarding civilisation that is presented by Nordenskjöld as inhumane. This contradiction on civilisational discourse, noted by Nordenskjöld as I will demonstrate, represents a form of in-between[9] in which the antagonist portrayal of civilisation and indigenousness is questioned, thus showing the mediator role of Nordenskjöld in the border zone.

The photographs will be analysed from a socio-semiotic perspective using the work of authors such as Barthes, Román Gubern and Umberto Eco,[10] also considering applications of their concepts in the field of the photographic representation of Indigenous peoples.[11]

---

8. Pratt, *Imperial Eyes*, 7.

9. H.K. Bhabha, "Culture's In-Between," in *Questions of Cultural Identity*, eds. S. Hall and P. du Gay (London: Sage 1996), 53–60. I will return to this concept in more detail when it is appropriate for the analysis.

10. See R. Barthes, "Le message photographique," *Communications* 1 (1961): 127–138; R. Barthes, "Rhétorique de l'image," *Communications* 4 (1964): 40–51; R. Gubern, *La mirada opulenta*, 3rd edition (Barcelona: G. Gili, 1994) and U. Eco, *La estructura ausente*, 3rd edition (Barcelona: Lumen, 1986).

Image 2. Nordenskjöld, "La terre de feu", 34

Le Tour du Monde magazine: Travel accounts in postcolonial contexts

There are records of at least four texts whose authorship is assigned to Nordenskjöld.[12] The first was published in Paris in the *Annales de Geographie* magazine. The second, shortly after the first, concerns an article published in Spanish (unknown translator) in *Actes de la Societé Scientifique du Chili* (Chilean Scientific Society) with the title, "Algunos datos sobre la parte austral del continente sud-americano, según estudios realizados por la comisión científica sueca."[13]

11. A. Azócar and J. Flores, "Así son, así somos: un texto, dos lecturas en la construcción de identidad." *Revista Comunicación y Medios* 13 (2002): 51–60; Azócar and Flores, "Fotografía de de capuchinos y anglicanos a principios del siglo XX: la escuela como instrumento de cristianización y chilenización." *Revista Memoria Americana* 14, (2006): 75–87; A. Azócar, *Fotografía Proindigenista. El discurso de Gustavo Milet sobre los mapuches* (Temuco: Ediciones Universidad de La Frontera, 2005).

12. O. Nordenskjöld, "L'expédition suédoise à la terre de feu 1895–97," *Annales de Géographie*, 28 (1897): 347–56; O. Nordenskjöld, "Algunos datos sobre la parte austral del continente sud-americano según estudios hechos por la comisión científica sueca," *Actes de la Société Scientifique du Chili*, 7: (1897): 157–68; O. Nordenskjöld, "La terre de feu," in *Le Tour du Monde, nouveau journal des voyages*. Nouvelle série, 8e année, (1902): 13–60; O. Nordenskjöld, *Viaje al Polo Sur* [English translation: "Translation from Swedish by Roberto Ragazzoni"] (Barcelona: Maucci, 1905).

13. English translation: "Some information about the southern part of the South American continent, according to studies made by the Swedish scientific commission."

Nordenskjöld's third text, which includes the photographs analysed in this chapter, is a selection and summary of notes written by the Swede. But this selection was made and translated into French by Charles Rabot,[14] and published in the Parisian magazine, *Le Tour du Monde*, under the concise title of "La Terre de Feu" (1902). Finally, we have a text in which Nordenskjöld talks about the long desired "Trip to the South Pole" (1905).[15] This text allows a better comprehension of the original purpose of Nordenskjöld's expedition to Patagonia: becoming the first Swedish mission to reach what at the time, from a European perspective, was one of the last geographical unknowns.[16]

The role of Charles Rabot as editor-commentator-translator and the circulation of Nordenskjöld's work in the French public, shows the relevance of the travel as a way of cultural transfer. The travelling of ideas written by a Swedish in Patagonia and published for a narrow scientific audience in the *Annales de Géographie* and for a larger public in *Le Tour du Monde* magazine, shows several networks that reveal the importance of this kind of texts.

Based on this information, it is clear that the texts of Nordenskjöld's work should be of great interest beyond his complex and somewhat frustrated expedition to the South Pole. The text here analysed ("La Terre de Feu") reveals a detailed geographical and geological description (the fundamental disciplines of Nordenskjöld), but also a racial, scientific and ethnographic narrative of the period, which is a result of meeting Selk'nam, Yaghan and Aónikenk Indigenous peoples, in particular.[17] In the same way, the interest in the geographical boundary issues provides us with examples of the imperial-colonial context,[18] which is

14. Although Rabot acted as an editor in this selection process, in which the text could be catalogued under the phenomenon of travel writing – where encyclopaedist erudition (scientific writing) is linked with entertainment and adventure (sentimental writing). In this sense, as a selector, summariser and commentator, Rabot himself could be considered as a "ghost writer." Regarding this, see Pratt, *Imperial Eyes*, 85–7.

15. We have the version of this text in Spanish and French, the one referenced here is the Spanish from 1905, which we consider as quite early considering the stay in Antarctica was between 1901 and 1903.

16. See E. Cruwys, "Antarctica," in *Literature of Travel and Exploration: An Encyclopedia*, ed. Jennifer Speake (New York: Routledge, 2013), 26–9.

17. In Sweden, the development of racial biology in the nineteenth and early twentieth centuries could provide a context for Nordenskjöld's thinking. See for example, J. Rogers and M. Nelson, "Lapps, finns, gypsies, jews, and idiots? Modernity and the use of statistical categories in Sweden," *Annales de démographie historique* 105, no. 1 (2003): 61–79.

18. E. Hobsbawm, *La era del imperio 1875–1914*, 6th edition (Buenos Aires: Editorial Crítica), 2009.

developed in the framework of building the sovereignty of Chile and Argentina in this period.

However, the text considered here was not published in a political, geographical or ethnographical review, but in a magazine addressed to a broader public. Travel magazines such as *Le Tour du Monde* appeared in this context of colonial expansion, as the globe was in the process of rediscovery, expansion and re-appropriation, with travel and the travellers/explorers becoming a key element in what would be a new articulation of territories and, ultimately, a new geopolitical conception of the world. This would ultimately lead to a "rediscovery of America,"[19] or from a more critical perspective, its invention.[20]

In this way, *Le Tour du Monde* represents one form of media that was used to broadcast the governing expansionist/colonialist ideology and "travel culture,"[21] inserted, at the same time, into the "colonial culture" that floods the thinking of the nineteenth century and drives the dreams, desires and fears of European society.[22] Magazines such as *Le Tour du Monde* and the stories it showcased are found "at the origin of the representations of space that will encourage travellers, and that will push them, eventually, to travel."[23] In the case of Latin America, the booming growth of bourgeois elites, consolidated after the wars of independence by the formation of nation-states, reproduced the colonial dynamics of internal

19. M. Huerta, "Le voyage aux Amériques et les revues savantes françaises au XIXe siècle," in *A la redécouverte des Amériques. Les voyageurs européens au siècle des indépendances*, eds. Bertrand Michel et Vidal Laurent (Toulouse: Presses Universitaires du Mirail, 2002), 73–93.

20. E. O'Gorman, *The Invention of America. An Inquiry into the Historical Nature of the New World and the Meaning of Its History* (Indiana: Indiana University Press, 1961); E. Dussel, *The Invention of the Americas. Eclipse of "the other" and the Myth of Modernity* (New York: Continuum, 1995). The position held by these authors is more critical because it assumes that the idea of the discovery was only valid from the European perspective, which makes invisible the fact that the "American" indigenous population knew the place where they lived and had a name for it in their own languages (Abya-yala, Tawantinsuyu, Anáhuac, among others). Thus, it is apparent that the nomination process does not have an objective or positivist reality but is constructed discursively. These themes have also been pointed out from the decolonial perspective, see W. Mignolo, *Historias locales/diseños locales. Colonialidad, conocimientos subalternos y pensamiento fronterizo* (Madrid: ed. Akal, 2003); W. Mignolo, *The Idea of Latin America* (Oxford: Blackwell, 2005), which studies the "global south".

21. S. Venayre, "Pour une histoire culturelle du voyage au XIX siècle," *Sociétés et Représentations* 21, Le siècle du voyage (April 2006): 5–21.

22. P. Blanchard, S. Lemaire and N. Bancel, "La formation d'une culture coloniale en France, du temps des colonies à celui des guerres de mémoire", in *Culture Coloniale en France. De la Révolution française à nos jours*, eds. P. Blanchard, S. Lemaire and N. Bancel (Paris: CNRS, 2008), 11–64.

23. Venayre, "Pour une histoire culturelle," 9.

colonialism, subjecting the indigenous populations to an exclusion based on the dichotomy of civilisation/savagery dichotomy.[24]

Image 3. Front image, *Le Tour du Monde*

Image 3 illustrates the ideal of territorial expansion prevalent at the time, as well as the new knowledge about the indexed territories, the re-appropriation and rediscovery of the world. This image appears on the cover of *Le Tour du Monde* from 1895, a year in which the magazine experienced some graphical changes.

Nordenskjöld's writing follows the formal and content requirements of *Le Tour du Monde*, which was founded in 1860 as a journal (associated with Hachette press) that mixed encyclopaedist instruction with popular entertainment. In this way, it contributed to the spectacularising of a still little-known world, which was assumed to be full of wonders, more or less quaint. Its editor, Edouard Charton, was a renowned journal entrepreneur (also the founder of the *Magazin pittoresque* and *L'Illustration*), who achieved recognition as a man of letters steeped in the anonymised spirit of St Simonism with *Le Tour du Monde*.

Hence, postcolonial ideology, or the logic of coloniality, is put to work through a series of cultural devices that legitimise the act of civilising, among which appear travel writing and their accompanying images. Postcolonial discourse is then related to the production of knowledge by agents of imperial powers, such as travel writers, missionaries, merchants, ethnographers and scientists.[25] This production

24. J. Flores, "Europeos en la Araucanía. Los colonos del Budi a principios del siglo XX." *Revista Iberoamericana* 8 (2000): 313–29; J. Rabasa, "Postcolonialism," in *Dictionary of Latin American Cultural Studies*, eds. M. Szurmuk and R. Mckee (Gainesville: University Press of Florida, 2012), 252–8; P. Navarro, "La conquista de la memoria. La historiografía sobre la frontera sur Argentina durante el siglo XIX." *Revista Universum* 20, vol. I (2005): 88–111.

of knowledge legitimises the act of civilising by colonial powers, the consequences of which can still be seen even today.

In addition, this production of knowledge is related to the desire for information about a distant cultural reference, which is one of the situations addressed by Michel Espagne and Michael Werner regarding cultural transfer.[26] Thus, Nordenskjöld's description of the foreign territory and the Indigenous people to the Swedish and French audience could be considered an act of cultural transfer, particularly in the manner of imagology, which is important to any understanding of the representation of identity and alterity – or "us" and "the others" and its link with cultural studies.[27] In this regard, Petra Broomans reminds us that Espagne's work is important for an understanding of cultural transfer within the more general framework of cultural studies.[28] As Lawrence Grossberg states in this regard:

> Cultural studies explore the historical possibilities of transforming people's lives by trying to understand the relationships of power within which individual realities are constructed. That is, it seeks to understand not only the organizations of power but also the possibilities of survival, struggle, resistance, and change.[29]

Moreover, the interaction network operating between a Swedish traveller, a French magazine and Chilean/Argentinian Patagonia, could be considered as a

---

25. L. Chrisman, "Postcolonial Studies," in *New dictionary of the history of ideas*, ed., M.C. Horowitz, 1857–1859 (Farmington: Thomson Gale, 2005).

26. M. Espagne and M. Werner, "La construction d'une référence culturelle allemande en France: genèse et histoire (1750–1914)". *Annales. Histoire, Sciences Sociales*, 42/4 (1987), 969–92.

27. The concepts of Identity/Otherness will be used to define, from the cultural studies perspective, the relationship of us/others. While *identity* is associated with the identical, i.e., to the "sameness" which represents the "us," the idea of *alterity* is associated with the other, i.e., to the "otherness" that represents the "others." See S. Rabinovich, "Alterity," in *Dictionary of Latin American Cultural Studies*, eds. M. Szurmuk and R. Mckee (Gainesville: University Press of Florida, 2012), 17–21; N. Solórzano and C. Rivera, "Identity," in *Dictionary of Latin American Cultural Studies*, eds. M. Szurmuk and R. Mckee (Gainesville: University Press of Florida, 2012), 181–92; A. Drace-Francis, "Identity," in *Keywords for Travel Writing Studies*, eds. Charles Forsdick, Zoë Kinsley, Kathryn Walchester (London: Anthem, 2019), 125–6; S. Burton, "Self," in *Keywords for Travel Writing Studies*, eds. Charles Forsdick, Zoë Kinsley, Kathryn Walchester (London: Anthem, 2019), 217–9.

28. P. Broomans, "The Meta-Literary History of Cultural Transmitters and Forgotten Scholars in the Midst of Transnational Literary History," in *Cultural Transfer Reconsidered. Transnational Perspectives, Translation Processes, Scandinavian and Postcolonial Challenges*, eds. Steen Bill Jørgensen and Hans-Jürgen Lüsebrink, Approaches to Translation Studies Series, Volume 47 (Leiden: Brill, 2021), 64–87.

29. L. Grossberg, "Cultural Studies," in *New dictionary of the history of ideas*, ed. M.C. Horowitz, (Farmington: Thomson Gale, 2005), 519–24, 520.

form of *histoire croisée*, stressing the transnational dimension, in this case, of the journey and the images conveyed as a result of it.[30] However, current thought has taken a critical position regarding the transnational, understood as an international literary space, because of its Eurocentrism and focus on Western societies.[31]

Us, the others and the in-between: From the hierarchy to the ambivalence

Here, the distinction between "us" and the "others" is fundamentally based on the civilisation/savagery dichotomy, as it used to be employed in the past. This strategic essentialism, which is here criticised, pretended to place the assumption of undeniable differences between categories of humans as a strategy for the construction of subaltern groups.[32]

In this way, even though each image shows a group of representatives of the "European" and of the "Indigenous," what is intended is to actually make a clear distinction between these subjects by showing, as I will demonstrate, some kind of hierarchy between Europeans and Indigenous people which is, nevertheless, questioned in a complex process of liminal spaces' construction.

The hierarchy of the travel account can be seen, in iconographic terms, by how the image of the Europeans presents them in a self-assured pose and with an awareness that they are being photographed, it gives the impression that they feel self-confident in front of the camera. The Indigenous people, in contrast, have a posture that portrays a certain discomfort, which is even clearer on viewing their faces. One of them is clearly suffering from the cold, which is confirmed on visualising that he is the only one not wearing shoes, while wearing a blanket-like cloak around his shoulders and body.

In terms of dress (iconic level), which without a doubt gives the image its symbolic investiture (iconographic level), there is also a huge difference. It is apparent that the Indigenous people are wearing a kind of uniform that homogenises them, their clothing is very similar; all of them are wearing the same kind of hat, two have scarves, trousers, jackets, etc. Meanwhile, the Europeans appear dressed in similar but clearly differentiated attire (unlike the uniformity of the indigenous people). The Europeans also have headwear, but each one is different from the others. There is no production of homogenisation through their clothing.

---

30. M. Werner and B. Zimmermann, "Beyond Comparison: Histoire Croisée and the Challenge of Reflexivity," *History and Theory*, vol. 45, no. 1 (2006): 30–50.

31. See Broomans, "The Meta-Literary History," 70–74.

32. See in this regard some of the concepts deployed by Ashcroft et. al., such as "binarism", "essentialism", "savage/civilised", to name a few, in Ashcroft, *Post-colonial studies.*

The iconic texts will help us understand the images better, placing them in the linguistic context of the travel writing into which they are inserted. Image 1 is preceded by the following paragraph:

> Thanks to these circumstances [to the monetary contribution of a "patron"] Doctor Otto Nordenskjöld could gather two naturalists: one botanist, the engineer P. Dusen, and one zoologist, M. Axel Ohlin, both of whom had proven themselves in previous explorations in the most various of latitudes. Once the mission was established, each of the members set out on his own. A reunion was expected in Buenos Aires.[33]

It is clear that those represented in the iconic text are those who take part in the expedition, which is also recorded in the caption that says: "Le docteur Nordenskjöld et ses compagnons de voyage" ("Doctor Nordenskjöld and his travel companions"). The image is thus left anchored to the linguistic elements encompassing it and all possible polysemy is removed.

On understanding the text, it is clear that what is done in the image is the positioning of the European ego, which will reveal how important exploration is for the expansion of civilisation. The travellers appear to be represented as qualified for the journey and that is why they have the erect self-certain posture that distinguishes them. This positioning of the European ego and its qualification is even better understood on reading the previous page, where the following is stated regarding Nordenskjöld's expedition:

> A Scandinavian, more than any other, can make observations of the greatest importance about this region. Located in the Southern Hemisphere at a latitude corresponding to that of southern Sweden, it presents to the naturalist, knowing perfectly the lands and biological conditions of the northern extremes of Europe, the elements for a comparison of great interest between the northern territories of the old world and the southern lands of the new continent.[34]

---

33. "Grâce à ce concours le docteur Otto Nordenskjöld put s'adjoindre deux naturalistes : un botaniste, l'ingénieur P. Dusen, et un zoologiste, M. Axel Ohlin, l'un et l'autre ayant fait leurs preuves dans de précédentes explorations sous les latitudes les plus diverses. La mission une fois constituée, chacun de ses membres se mit en route de son côté. Rendez-vous était pris à Buenos-Aires." Nordenskjöld, "La Terre de Feu," 14.

34. "Sur cette région, un Scandinave, plus que tout autre, peut rapporter des observations de la plus grande importance. Située dans l'hémisphère à une latitude correspondante à celle de la Suède méridionale, elle présente au naturaliste connaissant parfaitement les terrains et les conditions biologiques de l'extrémité septentrionale de l'Europe, les éléments d'un parallèle du plus haut intérêt entre les terres boréales de l'ancien monde et les terres australes du nouveau continent." Nordenskjöld, "La Terre de Feu," 13.

Thus, the Scandinavian is conceived as the ideal person to carry out the task of gaining knowledge because he already has knowledge of his own land, and the "new land" he is about to explore now is expected to be comparable, seen through the same lens and in this sense "the same thing." This principle of symmetry is a common element in intercultural approaches, where the other is made invisible by reducing it to the same.[35] Even if the parallelism is justified in part by pointing out the symmetrical location of the lands in both hemispheres, placing emphasis on the parallelism or comparison is already contradictory, because what will be compared must be seen through what is already known beforehand. Thus the travel account reduces the other to the same in this sense.

There is a presumption that the travellers need never get out of their own heads; that reaching these new lands only confirms what they already know. This is part of colonial thinking, described by Said: "To have such knowledge of such a thing is to dominate it, to have authority over it. And authority here means for 'us' to deny autonomy to 'it' – the Oriental country – since we know it and it exists, in a sense, *as* we know it."[36]

The domination of the space and of the other (the Indigenous in this case), is comprehended better on understanding the "civilisation" processes the Europeans deployed, where they sought to convert "barbarians" into "good savages."

It is helpful to visualise Image 2, anchored to the caption: *Fuegiens de la mission de Rio Grande* ("Fuegians from the Rio Grande mission"). In this way, it is clear that the Indigenous people represented are part of a civilising effort, characterised by the work of religious missionaries. Thus, the fact that the Indigenous people are uniformed, homogenised, with clothing that is very similar to the Europeans (but in difference to them) with a submissive, somewhat resigned posture, is relevant. In this sense, the story and the images themselves are transformed into cultural devices focused on the reproduction of "civilisation," where the other is contained and represented through dominant structures. The iconic representations form part of what, for Said, are representations of an imaginary geography

35. F. Hartog, "Les Scythes imaginaires: espace et nomadisme," *Annales. Économies, Sociétés, Civilisations.* 34e année, 6 (1979): 1137–54, 1139.

36. Said, *Orientalism*, 33. For more contemporary studies regarding the importance of Said for the discussion between travel writing and postcolonial, see: R. Clarke, "Toward a Genealogy of Postcolonial Travel Writing: An Introduction," in *The Cambridge Companion to Postcolonial Travel Writing*, ed. R. Clarke, (New York: Cambridge University Press, 2018), 1–18; J. Edwards, "Postcolonial travel writing and postcolonial theory," in *The Cambridge Companion to Postcolonial Travel Writing*, ed. R. Clarke, (New York: Cambridge University Press, 2018), 19–32; Lindsay, C. "Travel writing and Postcolonial studies," in *The Routledge Companion to Travel Writing*, ed. C. Thompson, (New York: Routledge, 2015), 25–34. Edwards, J. & Graulund, R. *Postcolonial Travel Writing: Critical Explorations*. (Basingstoke: Palgrave Macmillan, 2010).

and of the others that inhabit that geography. They produce a conceptualisation of the "territory of the others," which corresponds to "distinctive objects that, although they seem to objectively exist, only have a fictitious reality."[37]

This is all understood better on reading the accompanying text that surrounds the image of the Indigenous people from the religious mission:

> These Fuegians, far from being a race without intelligence and inferior, are on the contrary, superior to many primitives. Proof of their ingenuity is revealed in their hunting and fishing tools; no less compelling from this point of view is the ease with which they raise themselves in their contact with Europeans ... after a stay of six months in a white man's house, [a young Indian] could express himself in English and Spanish. This attentive, hard-working, punctual fellow was a perfect servant, such as those no longer found in civilised countries.[38]

Here, the Indigenous person, in the civilising process, is absorbed by the European-Self while at the same time becoming equal to the third person he/she. The Indigenous is associated with the European by the fact of "being equal," "the same," and on being equal, the Indigenous person will show the same characteristics and attitudes as the European (hard-working, timeliness, etc.). Assimilation works in this way, through a lens of equality, although evidently this is an equality only in name, since the Indigenous person becomes a "perfect servant," a perfect subordinate to the European.

Therefore, we can say that the European ego never abandons its Eurocentric view, and when the other is conceived as "equal" to them, it is because they find in the Indigenous person the same traits as the European: hard-working, timeliness, intelligence.

> Our traveller found another young Fuegian who spoke German, English and Spanish. After a few months of learning, the Indians become excellent cooks or housekeepers, whose services leave nothing to be desired. The adults understand their positions equally as quickly and are very hard-working.[39]

---

37. Said, *Orientalism*, 55.

38. "Ces Fuégiens, loin d'être une race inintelligente et inférieure, sont au contraire supérieurs à bien des primitifs. Témoin l'ingéniosité que décèlent leurs engins de chasse et de pêche; non moins probante à cet égard est la facilité avec laquelle ils s'élèvent au contact des Européens. (...) après un séjour de six mois chez un blanc, [a jeune indian] pouvait s'exprimer en anglais et en espagnol. Ce gamin attentif, laborieux, ponctuel, était un domestique parfait, comme on n'en trouve guère dans les pays civilisés," Nordenskjöld, "La Terre de Feu," 34.

39. As Rabot comments and translates the text : "notre voyageur a rencontré un autre jeune Fuégien qui parlait l'allemand, l'anglais et l'espagnol. Après quelques mois d'apprentissage, des Indiennes deviennent d'excellentes cuisinières ou des femmes de chambre dont les services ne

Homi K. Bhabha proposes that there is a "non-differential concept of cultural time,"[40] where the European does not realise that they are imposing ethnocentric criteria that are not typical of the other culture. In other terms, a respectful account of the otherness, which respects the differences between cultures, must be aware not only of the geographical distance, but also the time distance. Evidently, this does not mean to consider the other as "primitive" but as living in a different cultural time. That is the way in which a differential concept of cultural time should be regarded; by considering the historical genealogies of violence, colonialism, and so on:

> The sharing of equality is genuinely intended, but only so long as we start from a historically congruent space [space-time?]; the recognition of difference is genuinely felt, but on terms that do not represent the historical genealogies, often postcolonial, that constitute the partial cultures of the minority.[41]

In the equalising-homogenisation, Nordenskjöld even applies his criteria of justice to report the mistreatment suffered by the Indigenous people. The fact that one of the Indigenous persons is shown as shoeless in the image, and with clear signs of suffering from the cold, is not by chance. However, perhaps this also has to do with a kind of reporting that is made through the iconic image and that is ratified in the text:

> If these Indians had been treated well by the first colonists, if they had been given civilisation in moderate doses that they could have assimilated progressively, they would have been useful for the development of the South American republics. Instead, these unfortunates had been the prey of greedy adventurous men, and everywhere they had been hunted and killed as if they were wild beasts.[42]

---

laissent rien à désirer. Les adultes apprennent également rapidement des métiers et sont de très bons travailleurs," Nordenskjöld, "La Terre de Feu," 34.

40. Bhabha, "Culture's In-Between," 56. For contemporary authors using this concept, even if not related to the travel writing perspective developed here, see: M. Fernandez et.al., "Crossing the border between postcolonial reality and the *outer world*: Translation and representation of the third space into a fourth space," *Culture, Language and Representation.* Vol. XXI (2019): 55–70; B. Dröscher, "Travesía, travestí y traducciones. Posiciones in-between en la nueva novela historiográfica de América Central" *Revista de Estudios Sociales*, 12 (2002): 81–89; K. Batchelor, "Third Spaces, Mimicry and Attention to Ambivalence," *The Translator*, 14:1 (2008): 51–70.

41. Bhabha, "Culture's In-Between," 56.

42. "Si ces Indiens avaient été bien traités par les premiers colons, si on leur avait apporté à dose modérée une civilisation qu`ils eussent pu s'assimiler progressivement, ils auraient constitué un organisme utile pour le développement de républiques sud-américaines. Au lieu de cela, ces malheureux ont été la proie d'aventuriers avides, et partout ont été traqués et tués comme des bêtes fauves," Nordenskjöld, "La Terre de Feu," 34–35.

Here, Nordenskjöld has no problem in identifying his fellow colonialist European men ("greedy adventurous men"), but then he depicts the Indigenous people as animals, at least as dehumanised (in the relation hunter/prey), stating that the Indigenous have been "prey" and "murdered as if they were wild beasts." However, a kind of new representation is seen here, where civilisation and those implementing it, the travellers or settlers, would no longer be those propagating the benefits of civilisation to all, but rather the perpetration of inhumane evils. This would be a kind of "border thinking" or an "in-between in the culture," where the representation escapes the previously defined civilisation/savagery duality.[43]

This third space, an in-between, possibly appears as a means of assimilation in the form of mimicry, an issue that is seen in the way the image of the Indigenous people attempts to reproduce the posture identified with the Europeans. In this sense, the mimicry of colonial difference is manifest, in the manner indicated by Bhabha (2004):

> colonial mimicry is the desire for a reformed, recognizable Other, *as a subject of a difference that is almost the same, but not quite*. Which is to say, that the discourse of mimicry is constructed around an *ambivalence*; in order to be effective, mimicry must continually produce its slippage, its excess, its difference.[44]

In this sense, it is a border discourse that attempts to position the other through the illusion of homogeneity and that ultimately appears to fail, revealing the naivety of the attempt to construct a homogeneous identity. As such, travel as a physical movement also becomes a change or movement of ideological positions that includes a critique of the identity of the European, which can also be considered as a kind of cultural transfer.[45]

Thus, mimicry as it has been depicted above is a key for understanding the ambivalence which is proposed as a transit from the hierarchical perspective to a more undetermined one. In other words, ambivalence shows how the colonial discourse falls in self-contradiction since it appears as an exploitative relation instead of a nurturing one. As a result, the idea of reproducing the colonial discourse or subject (*mimic* the coloniser) changes into a mockery of the coloniser, which is the original idea from Bhabha to understand the mimicry.[46]

---

43. H.K. Bhabha, "Culture's In-Between," 54.

44. H.K. Bhabha, *The Location of Culture*. (New York: Routledge, 2004), 122.

45. P. Broomans and J. Klok, "Thinking about Travelling Ideas on the Waves of Cultural Transfer," in *Travelling Ideas in the long Nineteenth Century on the Waves of Cultural Transfer*, eds. P. Broomans, J. Klok and E. Bijl (Groningen: Barkhuis, 2019), 9–28.

46. H.K. Bhabha. *The Location of Culture*, 121–131.

This ambivalence between mimicry and mockery has already been exposed by several scholars:

> The problem for colonial discourse is that it wants to produce compliant subjects who reproduce its assumptions, habits and values – that is, 'mimic' the coloniser. But instead it produces ambivalent subjects whose mimicry is never very far from mockery. Ambivalence describes this fluctuating relationship between mimicry and mockery, an ambivalence that is fundamentally unsettling to colonial dominance.[47]

The analyses of the travel account suggest that Nordenskjöld seems to be completely aware of the ambivalence of the colonial dominance because of the contradiction of the civilised principles and the methods used to pursuit them:

> The expeditions by Lista and Popper begun in 1886 from Argentina to explore Tierra del Fuego were a calamity for the natives. Lista hunted the natives like game, persecuted them to extremes to capture some of them and force them to guide the group (...) After Lista and Popper a horde of adventurers arrived, attracted by the thirst for gold. They were also harsh and cruel to the Fueguians (...) After the gold prospectors came the colonists. Now the final act of this long drama begins, of this battle of a race of inoffensive primitives that civilisation, cornering them without mercy, has transformed into bands of thieves and scoundrels. And this final act is terrible. To get rid of their neighbours the colonists do not even hesitate to use strychnine![48]

At this stage, it seems useful to recall the fact that the concept of ambivalence was part of the psychoanalysis background[49] and adapted by Homi K. Bhabha to refer to the simultaneous attraction and repulsion that describes the tensioned relationship between coloniser and colonised.

In this regard, it can be said that the repulsion characterises not only the traveller's (coloniser's) attitude with regard to the Indigenous people (colonised), but

---

47. Ashcroft, et. al. *Post-colonial studies*, 10.

48. "Les expéditions de Lista et de Popper, parties en 1886 de l'Argentine pour prospecter la Terre de Feu, ont été calamiteuses pour les indigènes. Lista chassait les naturels comme un gibier, les poursuivant à outrance pour s'emparer de l'un d'eux et l'obliger à guider ses gens. (...) Après Lista et Popper arrivèrent une foule d'aventuriers attirés par le soif de l'or. Eux aussi se montrèrent durs et cruels à l'égard des Fuégiens. (...) Après les orpailleurs vinrent les colons. Maintenant, commence le dernier acte de ce long drame, de cette lutte d'une race de primitifs inoffensifs que la civilisation, en les traquant sans merci, a transformés en bandes de pillards et de malandrins. Et il est terrible ce dernier acte. Pour se débarrasser de leur voisins, les colons en reculent même devant l'emploi de strychnine!" Nordenskjöld, "Le Tour du Monde," 35–6.

49. See R. Young, *Colonial Desire: Hybridity in Theory, Culture and Race.* (London: Routledge, 1995).

also the coloniser and civilisation itself. This explains why Nordenskjöld wrote about the drama of the Indigenous, saying that civilisation "corner[ed] them without mercy" and transformed them into "bands of thieves and scoundrels".

Once again, this is crucial to understand how, in Nordenskjold's travel account, the binary thinking that separates the civilised and the savage is completely inverted. Note in this sense that the mimic idea behind the colonial discourse (mimic the values and habits of the coloniser by the colonised) cannot occur if the colonisers are "greedy adventurous men", a "horde", and if they do not have the positive features that the civilised world supposed to have.

Nevertheless, still there is a mimicry operating but instead of copying the good attributes of the civilisation, the Indigenous people become a "bands of thieves and scoundrels" precisely because that is the example they have seen in the Europeans. Evidently, from the perspective in which mimicry describes the tension between mimic and mockery, this reversal in the aspects of the civilisation shows all the contradiction and ambivalence that has been proposed so far. Thus, the binary between savage/civilised is here questioned by the appearance of an ambiguous way of conceiving civilisation's values.

Therefore, this text from the beginning of the twentieth century, accounts for a dynamic that escapes the civilisation/savagery dichotomy (or Western/Eastern dichotomy according to Said). This is important considering the fact that civilisation/Westernisation today appears negatively characterised in the literature in an almost essentialist way. Inevitably, from this dichotomous and binary perspective, it appears that all contact between civilisation/savagery will be negative, and even all possibility of a relationship is rejected, generating a theoretical basis where the cultures appear as limited, closed and incompatible entities.

Thus, I would argue that in the images analysed, the classic dichotomy of civilisation / savagery is, to a certain extent, placed in doubt through the narration and the images used in Nordenskjöld's account. Although it cannot be said that savagery is vindicated as a form of civilisation, it is clear that Nordenskjöld questions the supposed "civilisation" of the colonists and travellers who have come into contact in an "uncivilised" manner with Indigenous alterity. In this sense, through the text it is possible to corroborate a theoretical aspect referring to the existence of "border effects" in the discourse work of identity-building.

Here, we can read a "right to signify from the periphery of the authorized power,"[50] where the signification through "contingency and contradictory conditions" is rewritten.[51] What is interesting in this case, is that this resignification does

50. Bhabha, "Culture's In-Between," 19.

51. Bhabha, "Culture's In-Between," 19.

not start from those who are minoritised, but rather from the European, who is the centre of power, who is the one who makes room for a new signification that accounts for the European contradiction and positions the "Indigenous minorities" (in the sense of the "periphery of power") as significations that must not necessarily be assimilated to the "barbaric."

In other words, identity and alterity relation, as demonstrated in the image and the travel account, is mediated; it does not belong to the realm of the autonomous subject, but rather to the social subject surrounded by forms of power and other inherent elements which are constituted in a non-essentialist conception of identity that depends on the circumstances. Moreover, it is not so much on universal precepts, but rather on the particularities experienced by the subjects. This is a topic that appears clearly in the analysis of Nordenskjöld's characterisation of other Europeans as savages and of the Indigenous people themselves as bilingual, hard-working, punctual, etc.

It is with respect to this non-essentialist conception that Hall speaks of an identity linked to "agency," in the sense that identity is "strategic and positional."[52] The conflictual lens through which the relationship between civilisation and savagery is conceived, it is not a determined condition and unavoidable product of the "negative" and "positive" essences. Rather, it is one of the possibilities among many others that will depend on the particular type of relations that arise between different identities.

However, in saying this, there is no intention to deny the structural power leading the colonial relations, but to show one case – from Nordenskjöld- in which this structural power is, at least, questioned with an inversion of the significances related to Indigenous and European peoples.

As I have tried to show, this structural condition was colonial ideology in Nordenskjöld's time. Even considering what has already been mentioned concerning his complaint about the Europeans for their lack of civilisation in their treatment of indigenous people, it is still true that the importance given to Indigenous people is related to their usefulness for work activities (some kind of "good savage"), where Europeans remain in a position of power and Indigenous peoples are subordinates.

This can be observed in the following quotations of Nordenskjöld account, where he refers to the religious missionaries as the "true civilisation" rather than the other characters already discussed (gold prospectors, adventurers, hunters, etc.):

52. Cf. S. Hall, "Who needs identity?," in *Questions of Cultural Identity*, eds. S. Hall and P. du Gay (London: Sage, 1996), 14–17.

> While the colonists continued their savage conquest of Tierra del Fuego, the religious missions carried out the work of proselytism and true civilization. In 1888, the Salesians founded a school (...) the monks have obtained excellent results. The children receive primary instruction and a manual education; the girls learn to embroider and they always execute the work given to them with great care. The adults are also taught different trades, such as butcher, sawyer, shepherd, brick-maker (...) the Salesians attempt to initiate their students in the beauties of the trombone and the trumpet. Music soothes the soul, as they say! Nevertheless, the young Onas become insensitive to these enchantments and they frequently escape to resume the nomadic existence, the route through the deserts, which despite the inconveniences, is the dream life for the primitives.[53]

The paradox, and another evidence of the ambivalence and weak border zone described by Nordenskjöld, is that even the "true civilisation" is resisted by the Indigenous. Besides, the missionary effort, as shown in the Image 2, also presents some kind of contradiction or ambivalence considering what has been already said regarding the discomfort posture of the indigenous, the lack of shoes, cold suffering, etc.

Conclusion

The iconic representations that have been analysed in this chapter in relation with the travel account, appear to be descriptions of the other eluding the typical binary of the colonial border, this justify the idea of a "weak border zone". In this sense the discourse of the "us" and "others" appears as an illusion: "The border between the Same and the Other is guarded by the illusion of pure identity, enclosed by the 'inner experience' in its zeal to define the I."[54]

---

53. "Tandis que les colons poursuivaient cette conquête sauvage de la Terre de Feu, des missions religieuses entreprenaient une œuvre de prosélytisme et de véritable civilisation. En 1888, les Salésiens fondaient une école (...) les religieux ont obtenu d'excellents résultats. Les enfants reçoivent l'instruction primaire et une éducation manuelle; les filles apprennent la couture et exécutent toujours avec beaucoup de soin les travaux dont elles sont chargées. Aux adultes on enseigne également différents métiers, comme ceux de bûcheron, de berger, de scieur, de briquetier (...) les Salésiens essaient d'initier leurs élèves aux beautés du trombone et du cornet à piston. La musique adoucit les mœurs, dit-on! Néanmoins, les jeunes Onas demeurent insensibles à ses charmes et assez souvent s'échappent pour reprendre l'existence errante, la course à travers les déserts, qui, malgré ses privations, est pour les primitifs la vie rêvée." Nordenskjöld, "La Terre de Feu," 36.

54. Rabinovich, "Alterity," 20.

In practice, the existence of what has been described above as "border thinking" or a "border effect", where the representation escapes the civilisation/savagery dichotomy, results in an understanding of the postcolonial issue, and that of identity, beyond the binary scheme that opposes the coloniser and the colonised.[55]

However, and since there is still a colonial difference, the case of border thinking that has been characterised here responds to "weak border thinking" because "its emergence is not the product of the pain and fury of the disinherited themselves, but of those who are not disinherited and take their perspective."[56] In this sense of weak border thinking, the attitude of the traveller appears ambiguous or contradictory regarding the colonial violence that operates on the part of some Europeans and that Nordenskjöld himself condemns. It is contradictory because the traveller does not consider that his own presence in Patagonia is mediated by colonial violence, and it is ambiguous because the traveller moves between two worlds, the Indigenous and the European, but cannot, strictly speaking, leave his European perspective.

By moving between two worlds (between us and the others), Nordenskjöld acts as a cultural transmitter only in relation to a spatial distance between the Indigenous and the Europeans; Patagonia is far from the "civilised" imperial centres. However, he is not able to position himself in terms of a temporal distance, an issue that has been pointed out in connection with Bhabha's ideas on the "non-differential concept of cultural time," with which an ethnocentric and homogenising criterion is imposed. In this sense, a more holistic analysis of cultural mediation should consider these temporal issues.

References

Alvarsson, Jan-Åke, and Agüero, Oscar (Ed.). *Erland Nordenskjöld. Investigador y amigo del indígena.* Abya Yala: Quito, 1997.

Ashcroft, Bill, Griffiths, Gareth, and Tiffin, Helen. *Post-colonial studies. The key concepts* London: Routledge, 2007.

Azócar, Alonso. *Fotografía Proindigenista. El discurso de Gustavo Milet sobre los mapuches.* Temuco: Ediciones Universidad de La Frontera, 2005.

Azócar, Alonso and Flores, Jaime. "Así son, así somos: un texto, dos lecturas en la construcción de identidad." *Revista Comunicación y Medios* 13 (2002): 51–60.

55. See, in general, Rabasa, "Postcolonialism," 252. The idea of the postcolonial issue I mention can be understand, for the perspective developed so far, from the words of Rabasa: "In becoming conscious that the exercise of colonial power is not a matter of a mere binary schema that opposes the coloniser to the colonised resides the depth of postcolonial theory (...)," 257.

56. Mignolo, *Historias locales*, 28.

Azócar, Alonso. and Flores, Jaime. "Fotografía de capuchinos y anglicanos a principios del siglo xx: La escuela como instrumento de cristianización y Chilenización." *Revista Memoria Americana* 14 (2006): 75–87.

doi Barthes, Roland. "Le message photographique." *Communications* 1 (1961): 127–138.

doi Barthes, Roland. "Rhétorique de l'image." *Communications* 4 (1964): 40–51.

doi Batchelor, Kathryn. "Third Spaces, Mimicry and Attention to Ambivalence," *The Translator*, 14/1: 51–70. 2008.

Bhabha, Homi K. "Culture's In-Between." In *Questions of Cultural Identity*, edited by S. Hall and P. du Gay, 53–60. London: Sage, 1996.

Bhabha, Homi K. *The Location of Culture*. New York: Routledge, 2004.

Blanchard, Pascal, Lemaire, Sandrine, et Bancel, Nicolas. "La formation d'une culture coloniale en France, du temps des colonies à celui des guerres de mémoire", in *Culture Coloniale en France. De la Révolution française à nos jours*, eds. P. Blanchard, S. Lemaire and N. Bancel, 11–64. Paris: CNRS, 2008.

doi Broomans, Petra. "The Meta-Literary History of Cultural Transmitters and Forgotten Scholars in the Midst of Transnational Literary History." In *Cultural Transfer Reconsidered. Transnational Perspectives, Translation Processes, Scandinavian and Postcolonial Challenges*, edited by Steen Bill Jørgensen and Hans-Jürgen Lüsebrink, Approaches to Translation Studies Series, Volume 47, 64–87. Brill, 2021.

doi Broomans, Petra. and Klok, Janke. "Thinking about Travelling Ideas on the Waves of Cultural Transfer." In *Travelling Ideas in the Long Nineteenth Century on the Waves of Cultural Transfer*, edited by P. Broomans, J. Klok and E. Bijl, 9–28. Groningen: Barkhuis, 2019.

doi Burton, Stacy. "Self." In *Keywords for Travel Writing Studies*, edited by Charles Forsdick, Zoë Kinsley, Kathryn Walchester. London: Anthem, 2019.

Clarke, Robert. "Toward a Genealogy of Postcolonial Travel Writing: An Introduction." In Clarke, R. (ed.) 1–18. *The Cambridge companion to Postcolonial Travel Writing*. New York: Cambridge University Press, 2018.

Chrisman, Laura. "Postcolonial Studies." In *New dictionary of the history of ideas*, edited by M.C. Horowitz, 1857–1859. Farmington: Thomson Gale, 2005.

Cruwys, Elizabeth. "Antarctica." In *Literature of travel and exploration an encyclopedia*, edited by Jennifer Speake, 26–29. New York: Routledge, 2013.

doi Drace-Francis, Alex. "Identity." In *Keywords for Travel Writing Studies*, edited by Charles Forsdick, Zoë Kinsley, Kathryn Walchester. London: Anthem, 2019.

Dröscher, Barbara. "Travesia travesti y traducción posiciones in- between en la novela historiografía de america central," *Revista de Estudios Sociales*, 12: 81–89. 2002.

Dussel, Enrique. *The Invention of the Americas. Eclipse of "the other" and the Myth of Modernity*. New York: Continuum, 1995.

Eco, Umberto. *La estructura ausente*. 3rd edition. Barcelona: Lumen, 1986.

Edwards, Justine. "Postcolonial travel writing and postcolonial theory." In Clarke, R. (ed.) 19–32. *The Cambridge companion to Postcolonial Travel Writing*. New York: Cambridge University Press, 2018.

Edwards, Justine, and Graulund Rune. *Postcolonial Travel Writing: Critical Explorations*. Basingstoke: Palgrave Macmillan, 2010.

doi Espagne, Michel. and Werner, Michäel. "La construction d'une référence culturelle allemande en France: genèse et histoire (1750–1914)." *Annales. Histoire, Sciences Sociales* 42/4 (1987): 969–92.

doi Fernandez, María, Gloria Corpas, and Miriam Seghiri. "Crossing the border between postcolonial reality and the outer world: Translation and representation of the third space into a fourth space," *Culture, Language and Representation*. Vol. XXI: 55–70, 2019.

Flores, Jaime. "Europeos en la Araucanía. Los colonos del Budi a principios del siglo XX." *Revista Iberoamericana* 8 (2000): 313–29.

doi Fournier, Alex. "Expédition O. Nordenskjöld à la Terre de Feu." *Annales de Géographie*, 21 (1896): 341–2.

Grossberg, Lawrence. "Cultural Studies." In *New dictionary of the history of ideas*, edited by M.C. Horowitz, 519–24. Farmington: Thomson Gale, 2005.

Gubern, Román. *Iconic Messages in Mass Culture*. 1994.

Hall, Stuart. "Who needs identity?" In *Questions of Cultural Identity*, edited by S. Hall and P. du Gay. London: Sage, 1996.

doi Hartog, François. "Les Scythes imaginaires: espace et nomadisme." *Annales. Économies, Sociétés, Civilisations* 34e année, 6 (1979): 1137–54.

Hobsbawm, Eric. *La era del imperio 1875–1914*. 6th edition, Buenos Aires: Editorial Crítica, 2009.

doi Huerta, Mona. "Le voyage aux Amériques et les revues savantes françaises au XIXe siècle." In *A la redécouverte des Amériques. Les voyageurs européens au siècle des indépendances*, edited by Bertrand Michel and Vidal Laurent, 73–93. Toulouse: PUM (Presses Universitaires du Mirail), 2002.

Lindsay, Claire. "Travel writing and Postcolonial studies." In Thompson, C. (ed.) 25–34. *The Routledge Companion to Travel Writing*, New York: Routledge, 2015.

Mignolo, Walter. *Historias locales/diseños locales. Colonialidad, conocimientos subalternos y pensamiento fronterizo*. Madrid: ed. Akal, 2003.

Mignolo, Walter. *The Idea of Latin America*. Oxford: Blackwell, 2005.

doi Navarro, Pedro. "La conquista de la memoria. La historiografía sobre la frontera sur argentina durante el siglo XIX." *Revista Universum* 20, vol. I, (2005): 88–111.

doi Nordenskjöld, Otto. "L'expédition suédoise à la terre de feu 1895–97." *Annales de Géographie*, 28 (1897): 347–56.

Nordenskjöld, Otto. "Algunos datos sobre la parte austral del continente americano según estudios hechos por la comisión científica sueca." *Actes de la Société Scientifique du Chili*, 7 (1897b): 157–68.

Nordenskjöld, Otto. "La terre de feu." In *Le Tour du Monde, nouveau journal des voyages*. Nouvelle série, 8e année, (1902): 13–60. [Free translation].

Nordenskjöld, Otto. *Viaje al Polo Sur*. [Translation from Swedish by Roberto Ragazzoni], Barcelona: Maucci, 1905.

O'Gorman, Edmundo. *The invention of America. An Inquiry into the Historical Nature of the New World and the Meaning of Its History*. Indiana: Indiana University Press, 1961.

Pratt, Marie Louise. *Imperial Eyes. Travel writing and transculturation*. Routledge: New York, 2008.

Rabasa, José. "Postcolonialism." In *Dictionary of Latin American Cultural Studies*, edited by R. Mckee and M. Szurmuk. Gainesville: University Press of Florida, 2012.

Rabinovich, Silvana. "Alterity." In *Dictionary of Latin American Cultural Studies*, edited by R. Mckee and M. Szurmuk. Gainesville: University Press of Florida, 2012.

doi Rogers, John and Nelson, Marie. "Lapps, finns, gypsies, jews, and idiots? Modernity and the use of statistical categories in Sweden," *Annales de démographie historique* 105, no. 1 (2003): 61–69.

Said, Edward. *Orientalism*. London: Penguin, 2003 [1978].

Solórzano, Nohemy. and Rivera, Cristina. "Identity." In *Dictionary of Latin American Cultural Studies*, edited by R. Mckee and M. Szurmuk. Gainesville: University Press of Florida, 2012.

Spurr, David. *The Rhetoric of Empire. Colonial Discourse in Journalism, Travel Writing, and Imperial Administration*. Duke University Press: Durham, 1993. 2012.

doi Venayre, Silvayne. "Pour une histoire culturelle du voyage au XIX siècle." *Sociétés et Représentations* 21, Le siècle du voyage (April 2006): 5–21. [Free translation] (2006).

doi Werner, Michael. and Zimmermann, Benedict. "Beyond Comparison: Histoire Croisée and the Challenge of Reflexivity." *History and Theory* 45, no. 1 (2006): 30–50.

Young, Robert. *Colonial Desire: Hybridity in Theory, Culture and Race*. London, Routledge, 1995.

doi Youngs, Tim. "Border." In *Keywords for travel writing studies*, eds. Ch. Forsdick, Z. Kinsley, K. Walchester, 25–27, London: Anthem, 2019.

doi Zimmermann, Maurice. "Amérique du Sud. Rectifications de frontières." *Annales de Géographie*, 17 (1895): 518–9.

CHAPTER 5

Theatre as an engine for German-Swedish cultural transfer in the early twentieth century

Max Reinhardt's and Alexander Moissi's guest performances in Stockholm

Nina Brandau
Helmut Schmidt University Hamburg

It was not only literary exchange and travel writing that flourished between Germany and Scandinavia around 1900. The theatre sector was also influenced by increasing artistic mobility that facilitated the transfer of ideas on modern theatre across Europe. This chapter examines selected guest performances and directorships of the German theatre director Max Reinhardt and the actor Alexander Moissi in Stockholm between 1915 and 1921. Expanding the traditional idea of travel writing, the artists used guest performances as a means of allowing ideas to travel to a new national context. Using cultural transfer and mobility theories, the aim of this contribution is to explore how Reinhardt's and Moissi's mobile acts brought cultural and aesthetic ideas to Sweden and which factors – social, political, aesthetic as well as infrastructural – influenced the transfer process. This is done with a content analysis of newspaper articles as primary sources. With the help of these sources, I will illustrate that Reinhardt's and Moissi's visits to Sweden were only successful because of a well-developed German-Scandinavian network that had been built beforehand. It is argued that the artistic ideas that they transferred to Sweden were mainly affected by socio-cultural and infrastructural developments, while political ideologies were of secondary importance.

Keywords: cultural transfer, mobility, retransfer, travelling theatre, metropolitan culture, modern theatre, contact zone

https://doi.org/10.1075/fillm.20.05bra

> Up until today, itineraries and exchange processes remain the engine and heart of European theatre.[1]

The rise of new bourgeois classes and the development of a metropolitan culture in European cities facilitated a rise in international cultural exchange at the beginning of the twentieth century.[2] These developments also reinforced cultural transfer and artistic mobility between Germany and Scandinavia. The translations of Scandinavian works into German supported the reception of authors such as Henrik Ibsen and August Strindberg in Germany,[3] while Scandinavian artists found inspiration during visits to Germany, which influenced their work back home. Cultural exchange occurred in all sectors – including literature, painting and theatre. Nevertheless, thus far, research has mainly focused on the translation of literary works and their transmission between northern and continental Europe.[4]

A field that has been little studied in cultural transfer research between Germany and Scandinavia, so far, is the theatre, despite the quote above emphasising that mobility and transfer have always had a significant impact on European theatre. The several artistic and organisational levels that theatre consists of, and that exceed the textual basis of a book, make it an especially interesting phenomenon for transfer research and emphasise its potential for transmitting (aesthetic as well as political) ideas across borders. Firstly, there is the dramatic text and its author, including the inspirations and sources used in the writing. As depicted in several

1. "Bis heute sind Reisewege und Austauschprozesse der Motor und das Herzstück des europäischen Theaters," Peter W. Marx, *Max Reinhardt: vom bürgerlichen Theater zur metropolitanen Kultur* (Tübingen: Francke, 2006), 120.

2. Hubert van den Berg, Irmeli Hautamäki, Benedikt Hjartarson, Torben Jelsbak, Rikard Schönström, Per Stounbjerg, Tania Ørum, and Dorthe Aagesen, "Nordic Artists In The European Metropolises," in *A Cultural History of the Avant-Garde in the Nordic Countries 1900–1925* (Leiden, The Netherlands: Brill, 2012), 119, doi: https://doi.org/10.1163/9789401208918_009; Peter W. Marx, "Consuming the Canon: Theatre, Commodification and Social Mobility in Late Nineteenth-Century German Theatre," *Theatre Research International* 31, no. 2 (2006): 129–44.

3. Søren R. Fauth, Gísli Magnússon, and Peter Wasmus (eds), *Influx: der deutsch-skandinavische Kulturaustausch um 1900* (Würzburg: Königshausen & Neumann, 2014), 9.

4. Compare with e.g., Petra Broomans and Marta Ronne, "In the Vanguard of Cultural Transfer," in *In the Vanguard of Cultural Transfer: Cultural Transmitters and Authors in Peripheral Literary Fields*, ed. Petra Broomans and Marta Ronne (Groningen: Barkhuis, 2010), 1–12; Ester Jiresch, *Im Netzwerk der Kulturvermittlung: Sechs Autorinnen und ihre Bedeutung für die Verbreitung skandinavischer Literatur und Kultur in West- und Mitteleuropa um 1900* (Groningen: Barkhuis, 2013), Karin Hoff, Anna Sandberg, and Udo Schöning, "Einleitung," in *Literarische Transnationalität: kulturelle Dreiecksbeziehungen zwischen Skandinavien, Deutschland und Frankreich im 19. Jahrhundert*, ed. Karin Hoff, Anna Sandberg, and Udo Schöning (Würzburg: Königshausen & Neumann, 2015), 11–22.

chapters in this book, this textual basis can be influenced by different forms of travel, be it physical travel or the transfer of ideas (e.g. Broomans and Den Toonder in this volume). Secondly, there are the ideas, thoughts and actions of theatre directors, actors and others taking part in the production. Consequently, cultural transfer in theatre is not only dependent on single travel writers, but on a network of actors who are involved in a production. Thirdly, there is the performance itself, with the co-presence of the actors and the spectators, stage design, lighting and music. This layer emphasises that cultural transfer in theatre does not only take place through textual forms but is strongly bound to the physical presence of bodies. Thus, it reaches audiences beyond the literary interested public. Finally, the theatre institutions, as such, should be regarded as a decisive factor influencing the transfer of ideas and aesthetics. While publishing houses played a key role in travel writing,[5] the diverse infrastructural as well as personal settings of theatres as institutions had a strong impact on where a theatre production could travel to and how it unfolded in the specific space. Considering these multiple layers, I want to argue in this chapter that due to this network of human as well as (im)material components, theatre represented (and still represents) a very powerful way of transmitting ideas, politics and aesthetics across Europe.

In doing so, I will examine selected guest performances and directorships of the German theatre director Max Reinhardt as well as the performances of Alexander Moissi, one of the most famous actors in Reinhardt's company at the Deutsches Theater, which took place in Sweden between 1915 and 1921.[6] Using cultural transfer and mobility theories from Michel Espagne, Hans-Jürgen Lüsebrink, Stephen Greenblatt and Mary Louise Pratt, the aim is to identify how cultural and aesthetic ideas travelled to Sweden as a result of Reinhardt's and Moissi's guest appearances in Stockholm and what influence the social, political and aesthetic circumstances had on their work in Sweden. Referring to the theatrical context, this contribution seeks to show how the characteristics of traditional travel writing can also be applied to the analysis of a transnational theatre history, since research in this field has often primarily focused on national contexts.[7] A specific idea of the function of

5. Bill Bell, "The Market for Travel Writing," in *Handbook of British Travel Writing*, ed. Barbara Schaff (Berlin, Boston: De Gruyter, 2020), 125–141, https://doi.org/10.1515/9783110498974-008.

6. Even if both artists are not originally from Germany, they are considered to be German artists in this context, since they worked in Germany during the entire period of analysis and had a strong impact on German theatre development. An overview of all guest performances by Reinhardt and Moissi can be found in the appendix. Due to spatial limitations, only a select time frame will be analysed in this work.

7. Martina Groß, "Travel Literature and/as Transnational Theatre History – Beyond National Theatre Cultures," in *The Transnational in Literary Studies: Potential and Limitations of a*

Reinhardt and Moissi as travelling agents of culture is thereby revealed. The material examined in the case studies was transferred into another national context by these travelling agents, who transmitted their ideas on how to direct and perform theatre in a new environment. This act of mobility adds another dimension to our understanding of the transfer of modern theatre ideas and a reinterpretation of German and Scandinavian theatre texts that otherwise would not have been possible. Focussing on the role of Reinhardt and Moissi it becomes clear that, next to explorers (cf. the case of Nordenskjöld in this volume) and authors (cf. the case of Martinson in this volume), also theatre directors and actors play a decisive role in the exploration of travel writing in the Scandinavian sphere.

I will firstly place the artistic trips in a socio-cultural and political context to depict the circumstances under which the reinterpretation and transfer of the texts and performances occurred. Secondly, I will give an insight into the theoretical conceptualisation of cultural transfer and mobility research when analysing theatre performances from another era. Finally, I will illustrate how the single guest performances and productions of Reinhardt and Moissi transferred ideas about modern theatre to Sweden and how their work was influenced by the Scandinavian artistic environment as well as by international developments. To reconstruct the events of these times, Swedish newspaper reviews reporting on Reinhardt's and Moissi's Stockholm performances are analysed as primary sources. With the help of these sources, I will illustrate that Reinhardt's and Moissi's visits to Sweden were only successful because of a well-developed German-Scandinavian network that had been established beforehand and that socio-cultural and infrastructural developments mainly determined the artistic ideas that they transferred to Sweden, while political ideologies were of secondary importance.

It is important to mention that the focus here is more on Max Reinhardt as he had a strong influence on all artistic levels of theatre, due to his position as theatre director and his level of popularity in Europe. As an actor in Reinhardt's company, Moissi, in contrast, could mainly transfer his ideas by means of his performance on stage.

Internationalisation of and political impacts on theatre in the early twentieth century

In the first decade of the twentieth century, a metropolitan culture and new social classes emerged in many European cities which fostered international cultural

Concept, ed. Kai Wiegandt (Berlin, Boston: De Gruyter, 2020), 124–41, https://doi.org/10.1515/9783110688726-008.

exchange. Single artists who travelled abroad building up networks with their European colleagues, as well as entire artistic companies touring abroad, all influenced the cultural sector. This mobility and exchange of ideas across Europe was especially facilitated by new means of transportation and communication.[8] The installation of transnational railways allowed artists to travel to a greater extent than before and the invention of the telegraph supported rapid communication over greater distances.[9]

Around the turn of the century, a considerable number of Scandinavian artists was based in Berlin, studying the German performing arts, as well as the visual arts and writing, but also influencing Berlin's cultural life. The German capital belonged to the fastest-growing cities in Europe at that time and artistic life flourished – even during and after the First World War.[10] August Strindberg, Henrik Ibsen and Edvard Munch, among others, staged their plays for and exhibited their art to a German audience and contributed to the development of modern aesthetics in the German metropolis.[11] The results of Scandinavian-German exchanges were already perceivable in Reinhardt's work in Berlin before he and Moissi started to travel to Sweden. Staging Ibsen's drama, *Ghosts*, at the Deutsches Theater in 1906, Reinhardt collaborated with Edvard Munch to visualise new perspectives on the play.

August Strindberg also tried to introduce these modern waves into the Swedish theatre. In his post-inferno works, "the Swedish playwright strove for a theatre that he hoped might synthesise the objectivity of Naturalism with the spirituality of the Symbolists."[12] Therefore, he argued for simple stage settings, new lighting effects and music to stimulate an emotional theatre experience in the audience that went beyond pure reception of the dramatic text.[13] In his own

8. Bruce McConachie, "New Media Divide the Theatres of Print Culture, 1870–1930," in *Theatre Histories. An Introduction*, ed. Tobin Nellhaus et al. (London: Routledge, 2016), 10.

9. Orlando Figes, *The Europeans. Three lives and the making of a cosmopolitan culture* (New York: Metropolitan Books, 2019).

10. Jan Torsten Ahlstrand, "Berlin and the Swedish Avant-Garde – GAN, Nell Walden, Viking Eggeling, Axel Olson and Bengt Österblom," in *A Cultural History of the Avant-Garde in the Nordic Countries 1900–1925*, ed. Hubert van den Berg et al. (Amsterdam: Rodopi, 2012), 201, https://doi.org/10.1163/9789401208918_014.

11. Karin Hoff, Udo Schöning, and Frédéric Weinmann, "Einleitung/Introduction," in *Internationale Netzwerke: literarische und ästhetische Transfers im Dreieck Deutschland, Frankreich und Skandinavien zwischen 1870 und 1945*, ed. Karin Hoff, Udo Schöning, and Frédéric Weinmann (Würzburg: Königshausen & Neumann, 2016), 10.

12. McConachie, "New Media," 22.

13. Gösta M. Bergman, *Den moderna teaterns genombrott 1890–1925* (Stockholm: Bonniers, 1966), 267–74; Sverker Ek, "Brytningstid," in *1900-talets teater*, ed. Tomas Forser and Sven Åke Heed, Ny svensk teaterhistoria (Hedemora: Gidlund, 2007), 14–8.

Intima Teatern (1907–1910) in Stockholm, he attempted to enhance emotional effects by using a small theatre space, which strengthened the connection between the actors and the spectators.[14] He wrote several chamber plays for this modern spatial setting; however, these plays were very controversial and first favourably appreciated by the Swedish audience when Reinhardt staged them in Stockholm.[15]

These developments towards internationality in European theatre also had an influence on the audience's understanding of modern theatre. With the rise of performances in foreign languages the focus shifted from the literary text to theatrical interpretation. Language barriers did not seem to matter anymore, as the focus in modern theatre was on devices such as sound, light, space and corporeality which were used to create an overall impression – known as the *Gesamtkunstwerk*.[16] Indeed, the position of theatre as a "transmitter between national interests, international perspectives and intercultural connections"[17] was highly relevant around the turn of the nineteenth century, since it facilitated the exchange of ideas in modern aesthetics and avant-garde movements. The shift of focus from the dramatic text to the *Gesamtkunstwerk* created by a director led to a further internationalisation of theatre, since performances were no longer based on the work of one author from a particular nation but perceived as collective transborder phenomena created by international artists.[18]

Additionally, the political circumstances introduced by the First World War must also be considered in relation to Reinhardt's and Moissi's travels to Sweden. In general, it is important to emphasise that cultural life continued in European cities during the First World War and that culture's social function should not be underestimated. Theatre offers a platform to renegotiate cultural identity in times of social change, so it is not surprising that theatre changed significantly during the transformation of society at the beginning of the twentieth century.[19]

Even if Sweden maintained a neutral position throughout the war, the country's official stance towards Germany was benevolent neutrality.[20] Not only the artistic, but also the political connections between Sweden and Berlin were

14. Ek, "Brytningstid," 19–20.

15. Ek, "Brytningstid," 26.

16. Marx, *Max Reinhardt*, 123.

17. "Vermittlungsinstanz zwischen den nationalen Interessen, internationalen Perspektiven und interkulturellen Verbindungen," Hoff, Sandberg, and Schöning, "Einleitung," 16.

18. Hoff, Sandberg, and Schöning, "Einleitung," 16–7.

19. Suze van der Poll and Rob van der Zalm, "Introduction," in *Reconsidering National Plays in Europe*, ed. Suze van der Poll and Rob van der Zalm (Cham: Springer International Publishing, 2018), 8.

20. Inger Schuberth, *Schweden und das Deutsche Reich im Ersten Weltkrieg: die Aktivistenbewegung 1914–1918*, Bonner historische Forschungen 46 (Bonn: Röhrscheid, 1981), 16.

strong. Sweden, in contrast to its Scandinavian neighbours, had an "unusually intense orientation towards Berlin."[21] This led to a German interest in convincing Sweden to enter the war on its side and this was not only pursued on a political level but also by means of cultural propaganda. Theatre was an effective means to transport ideological content, since it was considered a "symbol of national culture, paragon of education and reserve of pure, ideal arts."[22] Especially with the rise of popular culture in the 1910s, a broader spectrum of society could be reached by theatre, as could its messages, since it was no longer predominantly attended by elites.[23] Referring to the thesis that German *culture* preceded Western *civilisation*, during the war, German intellectuals and the government aimed to strengthen this culture on a European level for the good of all Europeans.[24] This cultural policy, coupled with theatre's function as propaganda, was presumably one reason why the German foreign ministry played a large part in financing Reinhardt's guest performances in Stockholm in 1915.[25]

Despite the fact that theatre was influenced by the political events and ideologies, it is important to emphasise its capacity to maintain international exchange in times of nationalist countercurrents. International correspondence between key figures in literature and theatre was maintained during the First World War, providing a counter position to political developments.[26] Accordingly, the complex relationship between the cultural and the political domains during the First World War has a strong role to play in an analysis of cultural transfer and mobility across Europe.

---

21. Michael Jonas, *Scandinavia and the Great Powers in the First World War* (London: Bloomsbury Publishing PLC, 2019), 25.

22. "Symbol nationaler Kultur, Inbegriff der Bildung und Reservat reiner, idealer Kunst," Martin Baumeister, *Kriegstheater: Großstadt, Front und Massenkultur: 1914–1918*, Schriften der Bibliothek für Zeitgeschichte (Essen: Klartext-Verlag, 2005), 14.

23. Baumeister, *Kriegstheater*, 15–8.

24. Wolfgang J. Mommsen, "Einleitung: Die deutschen kulturellen Eliten im Ersten Weltkrieg," in *Kultur und Krieg: die Rolle der Intellektuellen, Künstler und Schriftsteller im Ersten Weltkrieg*, ed. Wolfgang J. Mommsen, Schriften des Historischen Kollegs 34 (München: R. Oldenbourg Verlag, 1996), 2; Peter Jelavich, "German Culture in the Great War," in *European Culture in the Great War: The Arts, Entertainment, and Propaganda, 1914–1918*, ed. Aviel Roshwald, Richard Stites, and Peter Jelavich, Studies in the Social and Cultural History of Modern Warfare 6 (Cambridge: Cambridge University Press, 1999), 45.

25. Schuberth, *Schweden und das Deutsche Reich im Ersten Weltkrieg*, 74 A 273.

26. Hoff, Schöning, and Weinmann, "Einleitung," 16.

Cultural transfer and mobility in theatre studies: An understudied phenomenon

Analysing cultural transfer and mobility during Reinhardt's and Moissi's visits to Stockholm under the circumstances mentioned above, the following questions must be raised on a conceptual level: Does the comparison of cultures lead to a hierarchisation? Are cultural areas limited by national borders, or have they ever been isolated from external influences? Does cultural exchange occur in a linear and one-dimensional manner? Answering these questions in the negative, the research on cultural transfer and mobility reaches beyond traditional comparative sciences.[27] Examining the transfer *between* rather than *of* cultures,[28] transfer research is not interested in comparing two cultures with each other but focuses on multidimensional networks within which cultural objects and ideas circulate. While nations have often been seen as cultural units in research undertaken in the twentieth century, and especially in theatre history,[29] cultural transfer theory emphasises that cultural areas are never and have never been restricted by politically or natively defined borders, and it strives for the creation of a transnational historiography.[30] The transfer of ideas has always been facilitated by acts of mobility that allowed ideas to travel from one cultural area to another.[31]

27. Jürgen Schriewer, "Problemdimensionen sozialwissenschaftlicher Komparistik," in *Vergleich und Transfer: Komparatistik in den Sozial-, Geschichts- und Kulturwissenschaften*, ed. Hartmut Kaelble and Jürgen Schriewer (Frankfurt: Campus Verlag, 2003), 36.

28. Christiane Eisenberg, "Kulturtransfer als historischer Prozess. Ein Beitrag zur Komparistik," in *Vergleich und Transfer: Komparatistik in den Sozial-, Geschichts- und Kulturwissenschaften*, ed. Hartmut Kaelble and Jürgen Schriewer (Frankfurt: Campus Verlag, 2003), 403; Quoted from: Michel Espagne and Michael Werner, "Deutsch-französischer Kulturtransfer als Forschungsgegenstand: eine Problemskizze," in *Transferts:les relations interculturelles dans l'espace franco-allemand (XVIIIe et XIXe siècle)* (Paris: Editions Recherche sur les civilisations, 1988).

29. Groß, "Travel Literature," 124.

30. Michel Espagne, "Jenseits der Komparatistik. Zur Methode der Erforschung von Kulturtransfers," in *Europäische Kulturzeitschriften um 1900 als Medien transnationaler und transdisziplinärer Wahrnehmung: Bericht über das Zweite Kolloquium der Kommission 'Europäische Jahrhundertwende – Literatur, Künste, Wissenschaften um 1900 in Grenzüberschreitender Wahrnehmung' (Göttingen, am 4. und 5. Oktober 2004)*, ed. Ulrich Mölk and Susanne Friede, vol. 273, Abhandlungen der Akademie der Wissenschaften zu Göttingen, Philologisch-Historische Klasse 3 (Göttingen: Vandenhoeck & Ruprecht, 2006), 13.

31. Stephen Greenblatt, "Cultural Mobility: An Introduction," in *Cultural Mobility*, ed. Stephen Greenblatt et al. (Cambridge: Cambridge University Press, 2009), 1–23, https://doi.org/10.1017/CBO9780511804663.

Michel Espagne has established one key concept for cultural transfer theory in the field of literature and the arts.[32] His main argument is that every shift of a cultural object from one context to another necessarily causes a change in its meaning, such that a new interpretation is required. It is not only the meaning of the object that changes but the new cultural context is also influenced by the object.[33] This transformation may concern medial, formal, semantic as well as functional modifications.[34] It is important to conceptualise the transfer as a process and not as a static phenomenon.[35] Espagne considers cultural areas to be hybrid constructs where cultural transfer has always occurred. One should not only analyse how two cultures have influenced each other but rather should map different cultural contacts that occurred between diverse areas. Proposing their actual hybrid character, he makes clear that cultural areas are primarily human constructions that are framed by discourses on nationality and identity.[36] Based on this premise, it should become clear why this paper focuses on consecutive visits of the two artists to Sweden, since a certain discourse was built around them in the course of the years they returned to Sweden.

Hans-Jürgen Lüsebrink's concept of cultural transfer builds on Espagne's thought and presents approaches to categorising and interpreting different processes in cultural transfer.[37] The categories range from simple adaptations that attempt to be as close to the original as possible, to the productive reception of transferred texts or discourses. Productive reception describes the creative adaptation of cultural texts or practices, which does not strive to imitate the transferred form but interprets it in the context of the target culture.[38] This kind of transfer is of special interest, since it reveals where the cultural lines blur and how dynamic the transfer processes are. The concept of "productive reception" will be the focus of the analysis here in interpreting how Reinhardt and Moissi crossed

32. Michel Espagne, "Der Begriff Kulturtransfer," in *Internationale Netzwerke: literarische und ästhetische Transfers im Dreieck Deutschland, Frankreich und Skandinavien zwischen 1870 und 1945*, ed. Karin Hoff, Udo Schöning, and Frédéric Weinmann (Würzburg: Königshausen & Neumann, 2016), 45–56; Espagne, "Jenseits der Komparatistik."

33. Espagne, "Jenseits der Komparatistik," 15.

34. Hoff, Schöning, and Weinmann, "Einleitung," 8.

35. Espagne, "Jenseits der Komparatistik," 15.

36. Espagne, "Der Begriff Kulturtransfer," 48 and 53–54.

37. Hans-Jürgen Lüsebrink, "Kulturtransfer," in *Interkulturelle Kommunikation: Interaktion, Fremdwahrnehmung, Kulturtransfer*, ed. Hans-Jürgen Lüsebrink, Christiane Solte-Gresser, and Manfred Schmeling, 4th revised and extended edition (Stuttgart: J.B. Metzler Verlag, 2016), 143–88.

38. Lüsebrink, "Kulturtransfer," 150–53.

cultural lines and created new forms of theatre with their performances and productions in Sweden.

Espagne's and Lüsebrink's approaches make clear that successful transfer research can only work if historical, political and social circumstances are taken into account.[39] Especially in times of politico-military expansion and hegemony, political interests have a strong impact on cultural transfer.[40] In the context of this research, in which two German artists physically travelled to Sweden to perform on Stockholm's stages, the theories of cultural transfer need to be set in the context of mobility studies. The physical movement of Reinhardt and Moissi is a significant determinant of the nature of the transfer. This physical mobility can be regarded as especially important when analysing transfers in theatre, since this artistic form is always bound to the corporeal presence of the actor and not only to the dramatic text, especially in a modern understanding of it. Here, we can make use of Stephen Greenblatt's theory of cultural mobility.

In addition to the historical, political and social circumstances, as described by Espagne and Lüsebrink, Greenblatt suggests taking the infrastructural and physical conditions into account to properly understand the way ideas travel.[41] Referring to Mary Louise Pratt, he underlines the crucial role of "contact zones" for the understanding of cultural mobility. Pratt conceptualises contact zones in opposition to homogenous communities. While the idea of communities is informed by a similar understanding of certain cultural norms, the contact zone is defined by the clash of different cultures and, consequently, by power imbalances.[42] According to Greenblatt, "certain places [in a cultural area] are characteristically set apart from inter-cultural contact; [while] others are deliberately made open."[43] When researching cultural transfer, we thus need to pay attention to the institutions and actors who facilitate cultural contact but also to the *blind spots* that were overlooked, intentionally or structurally. This theory especially emphasises the distinction between centre and periphery. Metropolises such as Stockholm or Berlin, with their well-established theatre institutions, offered several contact zones for the travelling artists, while the periphery was not visited by the artists and thus missed out on such cultural contact.[44] While other contribu-

39. Broomans and Ronne, "In the Vanguard of Cultural Transfer," 11; Espagne, "Der Begriff Kulturtransfer," 45; Lüsebrink, "Kulturtransfer," 143.

40. Lüsebrink, "Kulturtransfer," 156.

41. Stephen Greenblatt, "A Mobility Studies Manifesto," in *Cultural Mobility*, ed. Stephen Greenblatt et al. (Cambridge: Cambridge University Press, 2009), 250–3, https://doi.org/10.1017/CBO9780511804663.

42. Mary Louise Pratt, "Arts of the Contact Zone," *Profession*, 1991, 33–40.

43. Greenblatt, "A Mobility Studies Manifesto," 251.

tions focus on the temporal (e.g. Sandrock in this volume) or mystical dimensions (e.g. Yuan in this volume) of contact zones, in this case their spatial dimension is of relevance.

How can we analyse the above-mentioned social, historical, political and infrastructural factors in Reinhardt's and Moissi's performances when they took place more than 100 years ago? Every study that attempts to reconstruct a theatre event that took place in the past faces the problem of the uniqueness and ephemerality of all performance. In contrast to literature, theatre is always bound to a certain space and time.[45] The interaction between the actors and the audience, the use of lighting, colour, sound, movement and corporeality, as well as the general atmosphere in the theatre space, are perceived differently by every individual and can never be entirely reconstructed.[46] For the analysis of a theatrical event that happened in a time before video recording, theatre studies must rely on newspaper reviews, reports from the audience, prompt books and, in some lucky cases, photography, to reconstruct the performances as accurately as possible. However, due to the very individual nature of the experience of theatre performances, there will always be blind spots and biases in the reviews of performances.

Erika Fischer-Lichte offers a useful approach to analysing theatre productions. In her structural analysis, she does not use the dramatic text as a basis of analysis but focuses on the actual theatre event. This is especially useful when the linguistic level is not the focus in a performance, as in the case of Reinhardt's and Moissi's productions.[47] The productions can be divided into their different levels of meaning, such as sound, lighting and physical presence, and no chronological analysis of the productions is needed. Since newspaper articles only give insights into certain points of a performance, a chronological reconstruction would not be possible with the set of sources available.

An interpretive content analysis is applied here to trace these different levels of meaning in the newspaper reviews selected. This method is used "when direct access to original sources is limited [...] such as when the events are in the past

---

44. Petra Broomans and Janke Klok, "Thinking about Travelling Ideas on the Waves of Cultural Transfer," in *Travelling Ideas in the Long Nineteenth Century. On the Waves of Cultural Transfer*, ed. Petra Broomans and Janke Klok (Groningen: Barkhuis, 2019), 14–15.

45. Rainer Maria Köppl, "Das Tun der Narren. Theaterwissenschaft und Theaterdokumentation," in *Theaterwissenschaft heute: eine Einführung*, ed. Renate Möhrmann and Matthias Müller (Berlin: Reimer, 1990), 357.

46. Köppl, "Das Tun der Narren," 357.

47. Christopher Balme, *Einführung in die Theaterwissenschaft*, 5th revised and extended edition (Berlin: Erich Schmidt Verlag, 2014), 98.

or when participants are unavailable."[48] This requires the analysis of not only the content of the sources but also the meaning behind it.[49] To do so, the analysis attempts to extract descriptions of modern theatre developments from the subjectively written reviews, on the one hand, while it looks for hints that offer information on the organisational and social circumstances of the plays, on the other hand. The findings from the content analysis will subsequently be considered in relation to the socio-cultural and political context that was outlined above.

Nationality vs. Internationality: Cultural transfer in Reinhardt's and Moissi's Stockholm productions

Analysing German guest performances in Stockholm during the First World War, the focus is mainly on Reinhardt's trips with the Deutsches Theater company. As Moissi was a soldier in the German army and became a prisoner-of-war in France in 1915, he could not join the Deutsches Theater company for its Swedish guest performances in 1915 and 1917. It appears that attempts to negotiate with the French government in 1915 to exchange the actor for a French general captured by the Germans failed.[50]

The example of Moissi shows that the First World War influenced Reinhardt's Stockholm productions on an organisational as well as political level. This supports Greenblatt's argument that infrastructural or organisational components of mobility always have an influence on the metaphorical mobility of ideas.[51] On the organisational level, Reinhardt had to make changes to the personnel, and not only because Moissi was missing. The actress, Lucie Höflich, was too afraid to cross the Baltic Sea due to the mines laid during the War, so she refused to travel.[52] These personnel changes directly influenced the selection of Reinhardt's Stockholm repertoire, because Moissi and Höflich played key roles in certain performances. On the political level, Reinhardt had to adapt his

48. James Drisko and Tina Maschi, *Content Analysis*, Pocket Guides to Social Work Research Methods (New York: Oxford University Press, 2015), 59, https://doi.org/10.1093/acprof:oso/9780190215491.001.0001.

49. Drisko and Maschi, *Content Analysis*, 59. This is especially helpful when analysing theatre reviews that contain the critics' subjective assessments, which have to be transferred into the context of theatre studies to determine which aesthetic traits lie behind the spectators' emotions.

50. Anonymous, "Storartat gästspel av Reinhardt i Stockholm," *Aftonbladet*, 14 September 1915, Svenska Dagstidningar.

51. Greenblatt, "A Mobility Studies Manifesto," 250.

52. Kaifas, "Reinhardts gästspel börjar 2 maj," *Svenska Dagbladet*, 27 April 1917, Svenska Dagstidningar.

repertoire according to diplomatic intentions. This was especially the case in 1915, when the guest performances of the Deutsches Theater were largely subsidised by the German foreign ministry.[53] As mentioned above, Sweden's benevolent neutrality towards Germany in the first year of war led the German government to attempt to convince Sweden to enter the war on its side. Thus, the political goals of the state-funded German guest performances must also be taken into account.

During the war, the theatres in Berlin had two strategies to adapt their repertoire. The first was to focus on national classics such as Schiller and Kleist. The second was concentration on historical works, mostly with a reference to Prussian military history.[54] A focus on German classics is also noticeable in Reinhardt's 1915 repertoire in Stockholm.[55] Apart from one Strindberg play, the repertoire only consisted of authors from the canon of German classical dramas. One could certainly contest this, as two works by Shakespeare were also staged. However, after Reinhardt conducted an extensive survey among politicians and cultural actors on the appropriateness of staging Shakespeare in times of war, the respondents came to the conclusion that Shakespeare could be included in the repertoire of German classics. Arguing that Shakespeare's work was only properly understood and interpreted by German intellectuals, there was a general perception in Germany that the English author belonged to the German repertoire.[56] Reinhardt's focus on the classics instead of plays enthusiastic about war, and the way he interpreted them shows that he made little attempt to whitewash the cruel events of war and did not support nationalistic tendencies.[57] Instead, he attempted to follow a neutral line to continue his successful theatre business.

Nevertheless, it is remarkable that in both 1915 and 1917, Lessing's comedy *Minna von Barnhelm* was staged. The play had made a comeback on Berlin's stages at the beginning of the war. Its reference to the Seven Years' War, its reinforcement of the honour of the German soldier and its negative representation of the French made it fit the nationalistic mood.[58] The play's comedic character made it even more suitable for a broader audience and for hidden propaganda in the context of the First World War. Reinhardt's *Minna von Barnhelm* was not only well accepted in Germany, but was also praised by the Swedish audience as "the

53. Schuberth, *Schweden und das Deutsche Reich im Ersten Weltkrieg*, 74 A 273.
54. Baumeister, *Kriegstheater*, 52.
55. See production schedule in Annex I.
56. Baumeister, *Kriegstheater*, 56.
57. Eva Krivanec, *Kriegsbühnen: Theater im Ersten Weltkrieg; Berlin, Lissabon, Paris und Wien*, 37 (Bielefeld: Transcript, 2012), 99–100.
58. Krivanec, *Kriegsbühnen*, 101.

ideal interaction"[59] and an "accomplished work of art."[60] The synthesis of the stage set with the ensemble of actors made it a perfect *Gesamtkunstwerk*. It is interesting that the reviews from 1915 refer to the play's traditional character and the need to stage it in the right "temporal and local setting."[61] Reinhardt apparently followed these traditional lines in his staging to the favour of his Swedish audience: "There was an accomplished style realised throughout the play, which was certainly grounded in tradition."[62]

The entirely positive reviews reveal that the Swedish audience was still accustomed to traditional forms of performance without experimentation.[63] Reinhardt's stage experiments and new interpretations of bodies and spaces often led to negative reviews when they were first staged, as the unknown was not immediately accepted.[64] We can therefore assume that he did not cross aesthetic borders with his *Minna* production and that he did not intend to present very modern theatre in this case. The play was presumably supposed to please the German state and thus to convey a certain image of German *culture* and the German war mentality. Judging the production to be "high theatre culture,"[65] Swedish critics acknowledged the supposed primary position of German *culture* in a Europe that was opposed to Western *civilisation*.[66]

Even if *Minna von Barnhelm* and the selection of other German classics in 1915 point to a nationalistic and political tendency, Reinhardt primarily focused on the productions' artistic value. This becomes especially clear when looking at his Shakespeare productions in the 1910s. Here, the use of new techniques made available due to industrialisation becomes evident. Reinhardt saw high aesthetic potential in the classic works of the past being renewed in the contemporary context.

---

59. "det idealiska samspelet," S. L-n., "Det tyska gästspelet. 'Minna von Barnhelm'," *Aftonbladet*, 8 May 1917, Svenska Dagstidningar.

60. "fulländat konstverk," Bo Bergman, "Den fjärde Reinhardtaftonen. 'Minna von Barnhelm' en stormande succès," *Dagens Nyheter*, 13 November 1915, Svenska Dagstidningar.

61. "tids- och lokalfärg," Bergman, "Den fjärde Reinhardtaftonen."

62. "Där fanns en genomförd stil över det hela, som helt visst bottnar i tradition," Ernst Didring, "'Minna von Barnhelm' hos Reinhardt," *Aftonbladet*, 13 November 1915, Svenska Dagstidningar.

63. Bergman, "Den fjärde Reinhardtaftonen. 'Minna von Barnhelm' en stormande succès."

64. Erika Fischer-Lichte, "Sinne und Sensationen. Wie Max Reinhardt Theater neu erfand," in *Max Reinhardt und das Deutsche Theater: Texte und Bilder aus Anlass des 100-jährigen Jubiläums seiner Direktion*, ed. Roland Koberg, Bernd Stegemann, and Henrike Reinhardt Thomsen, Blätter des Deutschen Theaters 2 (Berlin: Henschel, 2005), 20.

65. "hög teaterkultur," Didring, "'Minna von Barnhelm' hos Reinhardt."

66. Jelavich, "German Culture in the Great War," 45.

Using new forms of theatre in his productions, he transformed the classics into images adopted from the modernised world.[67]

Therefore, it is not surprising that the reviews of his Shakespeare productions mainly focus on the external appearance of the performances, including the stage sets, lighting and the organisation of space. The 1915 reviews of *Twelfth Night* concentrate on the use of a revolving stage, which was specially constructed for Reinhardt's guest productions on the stage of the Operan (Royal Opera).[68] This was a key device in modern stage development, with electricity used to make the changes of scene smoother and enhance the spectators' sense of illusion. Reinforcing the *Gesamtatmosphäre* of the production, the revolving stage was not only a practical tool but has to be considered a dramaturgical device.[69]

However, the desired effect was not attained in the production of *Twelfth Night*. As Daniel Fallström wrote in his review: "let's give 'die Drehbühne' the slip; it creaks way too much and does not create any sense of illusion."[70] On the one hand, this negative perception of modern effects may have been due to the badly equipped stage at the Operan. The construction of a revolving stage for Reinhardt's guest performances must have been very provisional. This fact underlines Greenblatt's theory that infrastructural limitations influencing an act of mobility have a direct impact on the transfer of ideas.[71] On the other hand, the negative perception of this modern theatre device shows that the Swedish audience was not accustomed to such new productions. As Fischer-Lichte emphasises, it is mostly through negative reviews that we obtain information on the use of unusual theatre forms.[72] The statement by Fallström suggesting that Reinhardt sometimes crossed the line,[73] gives us a hint that he experimented along aesthetic lines in his Shakespeare productions.

The aesthetic rather than political focus becomes even more clear in Reinhardt's 1917 repertoire. Including works by Gorki and Kotzebue, the repertoire shows little influence of or attention to the diplomatic dimension. Reinhardt had already staged Kotzebue's *Die Kleinstädter* at the Kammerspiele in 1915 to underline his preference for a theatre that did not glorify war.[74] On the one hand, the

67. Marx, *Max Reinhardt*, 61.

68. Anonymous, "Storartat gästspel av Reinhardt i Stockholm."

69. Marx, *Max Reinhardt*, 57.

70. "låt oss slipda [sic!] 'die Drehbühne'; den gnisslar allt för mycket och ger ingen illusion," Daniel Fallström, "Deutsches Theaters andra föreställning. 'Was ihr wollt' von Shakespeare," *Stockholms-Tidningen*, 11 November 1915, Svenska Dagstidningar.

71. Greenblatt, "A Mobility Studies Manifesto," 250.

72. Fischer-Lichte, "Sinne und Sensationen," 20.

73. Fallström, "Deutsches Theaters andra föreställning. 'Was ihr wollt' von Shakespeare."

guest performances of 1917 emphasise Reinhardt's apolitical attitude and his artistic focus, while, on the other hand, they show that the interest that the German government had in bringing Sweden over to their side had ended. In 1916/17, Sweden's benevolent neutrality towards Germany had gradually changed, moving towards the idea of neutrality as a morally superior position in the international context.[75] This led to Germany giving up its attempts to convince the Swedish government.

Reinhardt's primarily artistic intentions during his visits to Sweden are clear. Updating old classics through the use of modern technology, he brought new interpretations of Shakespeare's work to Sweden and introduced modern international artists such as Maxim Gorki to his Swedish audience. This finding supports the thesis that transnational exchange in the arts and literature was often maintained in times of war and crises even if nationalist narratives dominated.[76] Additionally, Reinhardt should always be seen as a businessman. Despite his assistant, Felix Hollaender, emphasising in an interview that Reinhardt's primary interest with the guest performances during the First World War was to bring his theatre to an international audience even in times of war, we should not underestimate the economic factors involved.[77] In the first year of the War, the theatres in Berlin had feared for their existence. They were unsure whether German society would still be interested in going to the theatre or if it was appropriate to perform at all.[78] Therefore, the guest performances in a politically neutral, conveniently situated country, with an interest in new theatre aesthetics, at least in Stockholm, offered some guarantee of economic success.

> My predecessor Otto Brahm dedicated his life to making Henrik Ibsen accessible to the German people [...] Another Scandinavian poet, an allied, really Faustian nature, August Strindberg [...], left lasting documents to the German theatre after his horrible mental struggles. A new generation of actors grew up with this Nordic repertoire. They liberated themselves from all foreign influences which the theatre, with its world repertoire, naturally conserved the longest.[79]

74. Baumeister, *Kriegstheater*, 57.

75. Jonas, *Scandinavia and the Great Powers in the First World War*, 32–4.

76. Hoff, Schöning, and Weinmann, "Einleitung/Introduction," 11.

77. Sylvester, "Omkring Reinhardt," *Aftonbladet*, 4 November 1915, Svenska Dagstidningar.

78. Baumeister, *Kriegstheater*, 27.

79. "Mein Vorgänger Otto Brahm hat es sich zur Lebensaufgabe gemacht, Henrik Ibsen dem deutschen Volke nahzubringen, [...] Ein anderer skandinavischer Dichter, eine stammverwandte, eine wahrhaft faustische Natur, August Strindberg, [...], hat von seinem furchtbaren geistigen Ringen bleibende Dokumente dem deutschen Theater hinterlassen.[...] In diesem nordischen Repertoire ist ein neues Geschlecht von Schauspielern groß geworden. Sie haben

This comment by Reinhardt reinforces the significance of Scandinavian authors for German theatre and for the development of modern theatre as such.

In his introduction to *Fröken Julie* in 1888, Strindberg had already described his idea of an intimate theatre that would present the actors' psychological sensibility to the audience.[80] Taking up this idea, Reinhardt founded the Kammerspiele, an intimate stage, at the Deutsches Theater in 1906.[81] In turn, inspired by this, Strindberg opened his Intima Teatern in Stockholm in 1907.[82] While Strindberg mainly realised the aesthetic idea by composing plays for such an intimate setting, Reinhardt accomplished this modern form of theatre as a stage director.

Kammerspiele at Deutsches Theater Berlin (Architekturmuseum der TU Berlin, Inv. Nr. TBS 024,12.)

sich von allem fremden Einfluß befreit, dem das Theater mit seinem Weltrepertoire natürlich am längsten konservierte," Max Reinhardt, *Ich bin nichts als ein Theatermann: Briefe, Reden, Aufsätze, Interviews, Gespräche, Auszüge aus Regiebüchern*, ed. Hugo Fetting (Berlin: Henschelverlag, 1989), 413.

80. Ek, "Brytningstid," 19.

81. Marx, *Max Reinhardt*, 55–81.

82. Bergman, *Den moderna teaterns genombrott*, 282.

Strindberg's chamber plays were not broadly acknowledged when they were initially staged at his Intima Teatern. It was only during the guest performance of Reinhardt's Deutsches Theater company that Strindberg's more recent works were accepted by the Swedish audience.[83] Reinhardt used additional modern theatre forms such as lighting, sound and space to productively reinterpret the Swedish author's works. In this way, he gave the Swedish audience the opportunity to approach the chamber plays in a new manner and access their meaning.

While the performance of *The Dance of Death* in 1915 was not sold out and barely reviewed in the newspapers, the performances of *The Ghost Sonata* and *The Pelican* in 1917 and 1920 are among the highlights of the guest performances. The Swedish theatre world had started to identify with its national artist, as this quote by R.G. Berg from 1920 reveals: "We showed that we value our own more than others and that we understand Strindberg better than Schiller or Goethe."[84] Reinhardt clearly improved Strindberg's standing in Swedish society, but how did he achieve this acknowledgement of Strindberg's chamber plays on an artistic level? How did the (re)transfer of the chamber plays into the Swedish context succeed?

Reinhardt and Strindberg had a very similar understanding of how intimate theatre should present the inner psychological events to the audience by the actors' use of strong facial and physical expressions. Even if Reinhardt placed additional focus on the external effects, the key element in all of his Strindberg productions was precisely a focus on the actors' bodies. However, he not only used technical devices to reveal the character's psychological interior, but primarily used the body and body-related means to transfer energy from the actors to the spectators, as Fischer-Lichte points out. This concerned the production of sensuality, which is essential in Reinhardt's stagings.[85] Conveying this sensuality from the stage to the audience leads to interaction with and the physical involvement of the spectators.[86]

When examining the reviews of his *Thunder in the Air* production of 1920, it becomes obvious that Reinhardt aimed to express this sensuality with the chamber plays. Daniel Fallström believed this play was not suitable for a stage production,

---

83. Freddie Rock, *Tradition och förnyelse: svensk dramatik och teater från 1914 till 1922* (Stockholm: Akademilitteratur, 1977), 28.

84. "Vi visade, att vi sätta mera värde på vårt eget än på andra och att vi förstå bättre Strindberg än Schiller och Goethe," Ruben Gustafsson Berg, "Reinhardts fredagspremiär. Strindbergs 'Pelikanen' framföres under extatiskt bifall," *Aftonbladet*, 11 December 1920, Svenska Dagstidningar.

85. Fischer-Lichte, "Sinne und Sensationen," 23.

86. Fischer-Lichte, "Sinne und Sensationen," 17.

since it lacked a dramatic plot.[87] Nevertheless, his review of the performance is constantly positive, with praise for Albert Bassermann's natural and lively way of acting, which allowed the spectators to forget that they were in a theatre and facilitated a connection between the actor and the audience.[88] It was this energy and atmosphere that Reinhardt's actors transferred to the audience which especially made his Strindberg productions so successful. The spectators felt they were part of the performance and in this way interpreted Strindberg's chamber plays through sensual experience, even if the dramatic text did not convey much meaning.

It is especially interesting that Reinhardt succeeded in conveying this sensuality in the auditorium of the Royal Opera, as Strindberg's chamber plays were initially written for the small space of the Intima Teatern, which had only 161 seats, intending to create proximity between the actors and the spectators.[89] The Royal Opera, on the other hand, had 1100 seats.[90] Aware that the audience would not be able to trace all of the facial expressions and gestures on such a big stage, the voice played an important role in Reinhardt's productions here, where it was used as a physical means, as it could still be perceived from the back rows. According to Fischer-Lichte, the voice not only manifests the physical presence of the actors on stage but also creates spatiality and a sound sphere (*Lautlichkeit)*.[91] It plays an essential role in the creation of atmosphere. The production of a sound is directly linked to a physical process through the actors. When an actor produces a sound, it occupies the space of the auditorium and makes spatiality itself perceivable. Arriving at the spectators, the sound instigates an inner process that makes it possible to physically and sensually experience the event on stage.[92] When Fallström reports on the production of *The Ghost Sonata* in 1917, he mentions that Reinhardt presented the "sonata" in the right tone,[93] thereby already emphasising the importance of sound on a meta level. Later in his review, he mentions Gertrud Eysoldt's (German actress) imitation of a parrot while performing the role of the mummy. That

87. Daniel Fallström, "Deutsches Theaters gästspel. Sista föreställningen: Strindbergs 'Oväder'," *Stockholms-Tidningen*, 13 December 1920, Svenskt Teatergalleri, Theaterwissenschaftliche Sammlung Universität zu Köln.

88. Fallström, "Deutsches Theaters gästspel. Strindbergs 'Oväder'."

89. Ek, "Brytningstid," 19.

90. "Vår historia," Kungliga Opera, accessed 20 May 2019, https://www.operan.se/om-operan/historik/.

91. Erika Fischer-Lichte, *Ästhetik des Performativen*, 1st ed. (Frankfurt am Main: Suhrkamp, [2010), 219.

92. Fischer-Lichte, *Ästhetik*, 226.

93. Daniel Fallström, "Deutsches Theaters gästspel. Strindbergs 'Spöksonaten' andra programmet," Stockholms-Tidningen, 4 May 1917, Svenskt Teatergalleri, Theaterwissenschaftliche Sammlung Universität zu Köln."

this part is even mentioned in the review shows the importance of such sounds for the spectators and the atmosphere. Reinhardt's focus on the atmospheric impact is summarised by Fallström: "One was deeply moved, bewitched, taken away by this fantastic vision on the edge of brilliance and madness."[94]

Reinhardt achieved two things in staging Strindberg's chamber plays in front of a Swedish audience. On the one hand, he created a new awareness of Strindberg's recent works, allowing the Swedish audience to gradually identify with a national author. On the other hand, he helped the Swedish audience develop a new sensibility for performance. Shifting the focus from an intellectual to a sensual theatre experience that makes the spectators part of the performance, the Swedish audience became aware of a modern way of perceiving theatre. Of course, these two factors are closely linked to each other. Taking Espagne's and Lüsebrink's theories into account, we might say that Reinhardt combined a text that originally stemmed from Sweden with theatre aesthetics that he had mainly developed in his theatre in Berlin to retransfer the work of art to Sweden. It was not Reinhardt's aim to stay as close to the original as possible, but rather to use productive reception.[95] Not surprisingly, the way he staged Strindberg's text had a big impact on how he was received in Sweden.

Additionally, this retransfer of a Swedish text shows how much mobility had played a part before Reinhardt's Stockholm visits. He was only able to reinterpret Strindberg's works due to the fact that the Swedish texts had already made their way to Germany some years before and had been translated into German. Greenblatt emphasises that mobility should always be seen in relation to rootedness and the fact that certain ideas develop in a specific context. This example of retransfer makes clear the close interconnection between rootedness and mobility.[96]

Turning now to Alexander Moissi, who also used the text of a Scandinavian author to present his performative skills to a Swedish audience. His Swedish premiere as Osvald in Henrik Ibsen's *Ghosts* in April 1921 joined a long German-Scandinavian performance history of this work. After the play was first staged in Augsburg, Germany, in 1886, the theatre director Otto Brahm opened his experimental stage, the Freie Bühne, in Berlin with *Ghosts* in 1889.[97] In doing

94. "Man satt fullständigt gripen, förhäxad, bergtagen av denna underliga fantastiska vision på gränsen mellan geni och galenskap," Fallström, "Deutsches Theaters gästspel. Strindbergs 'Spöksonaten.'"

95. Lüsebrink, "Kulturtransfer," 162–67.

96. Greenblatt, "A Mobility Studies Manifesto," 252.

97. Erika Fischer-Lichte, "Ibsen's Ghosts – A Play for All Theatre Concepts?: Some Remarks on Its Performance History in Germany," *Ibsen Studies* 7, no. 1 (June 2007): 62–5, https://doi.org/10.1080/15021860701464505.

so, he presented a new naturalistic style that was inspired by the Théâtre Libre in Paris.[98] In 1906, 17 years later, Reinhardt opened his Kammerspiele in Berlin with the same play. Producing an originally naturalist play, he continued a performance tradition but at the same time went beyond it with a new interpretation. Cooperating with the Norwegian artist Edvard Munch, who designed the stage set, Reinhardt emphasised yet another modern development in theatre. Munch did not follow a naturalistic line by presenting a Norwegian landscape on stage but visualised the inner psyches of the main characters, reinforcing the psychological effect of the play. He gave the spatial setting on stage an essential storytelling function, using strong colours such as red or yellow aiming for a captivating *Gesamtkunstwerk*.[99] Only with the support of the Norwegian artist was it possible for Reinhardt to overcome realistic imitation in an original naturalistic work.

Alexander Moissi initially played the role of Osvald in Reinhardt's 1906 production. Fifteen years later, he took part in Ernst Eklund's *Ghosts* production at the Blancheteatern in Stockholm. Of course, Ibsen's play had been performed on Swedish stages before, but this latter production was based on a new German-Scandinavian collaboration, where a new contact zone for cultural mobility was established. In line with Espagne's theory, this staging in a new context reveals aesthetic features and meanings that require reinterpretation.[100]

While Ibsen's *Ghosts* was a popular example of the socio-critical content on stage in the late nineteenth century,[101] this function of his work was far weaker in 1921. Rather than general social problems, the individual psyche was the focus in this production, using new theatrical devices to adapt the play to the contemporary context. Nevertheless, after the cruelties of the First World War, the critics of Eklund's *Ghost* not only saw the aesthetic achievements of the German actor Moissi as remarkable, but emphasised Ibsen's renewed social actuality. After the confusion of the War, the audience was expected to be able once again to concentrate on deep content and thoughts.[102] Ibsen gained a new relevance after the War, which had changed the audience's view on culture and theatre. However, when looking deeper into the 1921 production, it becomes clear that modern theatre

98. Bergman, *Den moderna teaterns genombrott*, 104.

99. Marx, *Max Reinhardt*, 69–70.

100. Espagne, "Der Begriff Kulturtransfer," 45.

101. Marx, *Max Reinhardt*, 65.

102. Ruben Gustafsson Berg, "Ibsens 'Gengångare' på Blancheteatern. Moissi som Osvald," *Aftonbladet*, 27 April 1921, Svenskt Teatergalleri, Theaterwissenschaftliche Sammlung Universität zu Köln.

aesthetics and a certain connection between the actor and the audience were of greater importance than the play's socio-critical concerns.

One important feature that fostered the development of modern theatre aesthetics in this production was the mixture of languages. While Moissi performed in German, his co-actors spoke Swedish. Even if a large part of Sweden's population was able to understand some German at that time, we can assume that the switch between two languages made it impossible for many of the spectators to follow the performance on a purely linguistic level. As a result, the meaning was not transferred to the audience by the dramatic text alone, but by the use of a body-centred performance style, including the detailed use of the voice and body movements. Several reviews emphasise Moissi's voice. As one reviewer describes it: a "dry voice with its deep wistfulness and sudden jarring metal sound."[103] He made his voice fit the character's mood and in this way transferred the meaning to the audience through his physical presence, which, according to Fischer-Lichte, includes the intonation and sound of an actor's voice.[104] In this way, the physical and emotional reception by the audience was prioritised over the linguistic and intellectual reception.

Even if the linguistic level was of secondary relevance, the fact that the languages were mixed also had an influence on the general meaning of the play. Espagne makes clear "that words are enrooted in semantic contexts and that the shift of a semantic context as a consequence of a translation always represents a new construction of meaning."[105] Thus, a play that was originally written in Norwegian and then performed in a German-Swedish version conveys a different meaning than in its original version.

It was not only Moissi's multifaceted use of voice and physical expression but also the performance of his co-actress, Karin Swanström, who played Fru Alving, that added to the aesthetic value of the production: "The actress showed a rich soul life, one forgot the constructed character in front of such a moving humanity."[106] The natural but detailed way in which they both performed their roles fostered the creation of a *Gesamtkunstwerk*, as their performance of the roles came together well, with no aesthetic breaks perceivable.

---

103. "torra röst med dess tunga vemod och plötsligt skärande metallklang," Einar Smith, "Moissi på Blancheteatern," *Svenska Dagbladet*, 27 April 1921, Svenska Dagstidningar.

104. Fischer-Lichte, "Sinne und Sensationen," 19.

105. "dass Begriffe in semantischen Kontexten verwurzelt sind und dass die Verschiebung des semantischen Kontextes in Folge der Übersetzung eine neue Sinnkonstruktion darstellt," Espagne, "Der Begriff Kulturtransfer," 54.

106. "Skådespelerskan avslöjade ett rikt, storsinnat själsliv, man glömde det konstruerade i rollen inför en så gripande mänsklighet," Smith, "Moissi på Blancheteatern."

Eklund efficiently used the proximity of the spectators and the actors provided by the small space at Blancheteatern. In line with the idea of an intimate theatre,[107] the audience was able to trace every movement and emotional expression occurring on stage. Since the reviews do not give any information about the décor at Blancheteatern, we can assume that the performances of Moissi and Swanström expressed a more intensive meaning than the stage settings. However, this intimate form of theatre also generated a different manner of performing by the actors.[108]

With Moissi, Eklund chose a very experienced actor who had learned how to adapt to a small stage at Reinhardt's Kammerspiele. It seems that the language differences between Moissi and the audience gave him even more opportunity to focus on his physical presence. While his natural, and not only beautiful, way of performing Osvald was criticised in Reinhardt's production of 1906,[109] he received only compliments from the Swedish critics. The reason for the German actor's success at the guest performance can, on the one hand, be attributed to his greater experience 15 years later. On the other hand, the language difference led to an alternative perception of his performing style by the Swedish audience. Yet another explanation could be that the Swedish audience in 1921 was already more used to intimate theatre and the modern form of performance than Reinhardt's audience had been in 1906 at the inauguration of the first intimate theatre in Berlin. In Stockholm, Strindberg's Intima Teatern had been operating between 1907 and 1910 and, subsequently, the Nya Intima Teatern, as well as other private theatres, had continued the tradition of a small theatre space in the 1910s. Therefore, at the beginning of the 1920s, the concept of an intimate theatre was already well established in Stockholm.

In summary, there were three new features central to the cooperation between Eklund and Moissi. The first was the use of multilingualism on stage and the way this supported a certain way of performing. This body-focused performance style that concentrated on pathos matched the spatial environment of the Blancheteatern well. The second was Moissi's experience performing Ibsen and in adapting his way of acting to an intimate setting. Contributing to this was the strong influence of Reinhardt during Moissi's years at the Deutsches Theater company. The third feature was the post-War spirit, which gave the play new social relevance, on the one hand, because the War had visualised the abyss of human misery on a new scale. On the other hand, while interest in *light* entertainment

107. Marx, *Max Reinhardt*, 72–81.

108. Marx, *Max Reinhardt*, 79.

109. Marx, *Max Reinhardt*, 79.

Intima Teater 1907 (Strindbergsmuseet, Stockholm)

had grown between 1914 and 1918, after the War, Ibsen was seen as an opportunity to bring more intellectual works back to the theatre.

It seems logical that one critic described the play in the following way: "It was at times entirely modern, at times completely 'strindbergskt', sometimes it evoked memories of former and even the oldest times."[110] Cooperation between a Swedish director and a German actor led to a new form of theatre. The many influences on the performance show how it profited from German-Swedish transfers that had already occurred and how it contributed to new transfers. Referring to Pratt, this example shows how fruitful a certain contact zone may be.[111] The institutional setting of the theatre opened the possibility for the cultural contact of a German artist with a Swedish director performing a Norwegian drama in several languages before a Swedish audience.

110. "Det var delvis rent modernt, föreföll stycketvis rent strindbergskt, delvis väckte det hågkomster från forna och de allra äldsta tider," Gustafsson Berg, "Ibsens 'Gengångare' på Blancheteatern. Moissi som Osvald."

111. Pratt, "Arts of the Contact Zone."

Conclusion

The transfers during the guest performances of Reinhardt and Moissi were strongly initiated and facilitated by aesthetic as well as social developments in Germany and Sweden during the first decades of the twentieth century. At the same time, the impact of certain people in their network on their success in Sweden should not be underestimated. All of these factors are only relevant due to one determining condition: mobility. The transfer of ideas was possible because Reinhardt and Moissi were able to travel to Sweden even during the First World War.

The relevance of Ibsen's and Strindberg's works for their success should also be emphasised here. Without the modern plays of the Scandinavian artists and Strindberg's primary thought of an intimate theatre, it would not have been possible for neither Reinhardt nor Moissi to be as successful as they were in the Scandinavian cultural area. It is likely that the standing of Strindberg's work would have also remained underestimated by the Swedish audience. This mutual benefit of the German-Swedish transfer supports the assumption of Hoff and colleagues that we cannot define Scandinavia as a peripheral area when analysing cultural transfers, neither in literature nor in theatre.[112] Many modern developments were significantly co-created by Scandinavian artists and not only implemented in the North by central Europeans. Similar contact zones for cultural exchange also existed in Berlin and the Nordic capitals.

Addressing the aesthetic elements of modern theatre that were transmitted, it is obvious that the elements that were focused on depended on the plays performed. While the production of Scandinavian plays focused on the actors' bodies, the Shakespeare productions experimented strongly with the stage design and lighting effects. This differentiation according to the styles of the plays is in line with Reinhardt's vision of three different stages. He called for a large stage to produce the known classics, a small one for the chamber plays of modern authors and a huge stage for mass productions such as occurred in Ancient Greece.[113] Focusing on the different theatrical elements, Reinhardt and Moissi supported an ongoing development in Swedish theatre that might have taken longer without their influence. Theatre turned from a text-bound, naturalistic form to one that brought the synthesis of the arts and the emotional effects on the spectators into focus.

Regarding the impact of political developments, it is clear that Reinhardt had to adapt to the First World War on an organisational level, but the transfer of the essential idea of modern theatre was not hampered by the political conflict between 1914 and 1918. Indeed, Reinhardt's Stockholm productions in 1915 and

112. Hoff, Schöning, and Weinmann, "Einleitung," 9.

113. Andreas Kotte, *Theatergeschichte: eine Einführung* (Köln: Böhlau, 2013), 297.

1917 showed how "the correspondence between the artists and intellectuals [...] often withdrew from political influence and countered it with a productive and contradictory exchange."[114]

The way that Reinhardt adapted his 1915 repertoire to please his sponsor, the German government, particularly emphasises his position as a businessman. Along with Reinhardt's artistic significance, it should never be forgotten that he also had to manage the financial side of his international tours. Since it was not possible to take his role as a businessman sufficiently into account in this work, it would be worthwhile undertaking further research on the financial aspects of his Stockholm trips to discover the role of artistic aims and to what extent his travels to Scandinavia brought financial advantage to his theatre business. In line with mobility studies, this organisational factor also needs to be taken into account to properly understand the transfer of ideas and the reasons for this transmission.

All in all, the analysis pointed to the potential that theatre has as a site of cultural transfer and the important role of mobility in this artistic field. The idea of the *Gesamtkunstwerk*, which underlines the direct effect that theatre has on the spectators, particularly points to the medium's potential to transfer certain aesthetics, ideologies and political ideas across borders and language barriers. Since theatre aims for a physical experience by the spectators, it can convey meaning even if the text is not understood by the audience. This international character of theatre shows the important function it can still have today in allowing ideas to travel to other countries, reinforcing international exchange.

References

Ahlstrand, Jan Torsten. "Berlin and the Swedish Avant-Garde – GAN, Nell Walden, Viking Eggeling, Axel Olson and Bengt Österblom." In *A Cultural History of the Avant-Garde in the Nordic Countries 1900–1925*, edited by Hubert van den Berg, Irmeli Hautamäki, Benedikt Hjartarson, Torben Jelsbak, Rikard Schönström, Per Stounbjerg, Tania Ørum, & Dorthe Aagesen, 201–27. Amsterdam: Rodopi, 2012.

Anonymous. "Storartat gästspel av Reinhardt i Stockholm." *Aftonbladet*. 14 September 1915. Svenska Dagstidningar.

Balme, Christopher. *Einführung in die Theaterwissenschaft*. 5th revised and extended edition. Berlin: Erich Schmidt Verlag, 2014.

Baumeister, Martin. *Kriegstheater: Großstadt, Front und Massenkultur: 1914–1918*. Schriften der Bibliothek für Zeitgeschichte. Essen: Klartext-Verlag, 2005.

114. "die Korrespondenzen zwischen den Künstlern und Intellektuellen [...], sich häufig politischer Vereinnahmung entziehen und ihr einen produktiven und politischen Verhärtungen widerstreitenden Austausch entgegensetzen können," Hoff, Schöning, and Weinmann, "Einleitung," 16.

doi Bell, Bill. "The Market for Travel Writing." In *Handbook of British Travel Writing*, edited by Barbara Schaff, 125–141. Berlin, Boston: De Gruyter, 2020.

Bergman, Bo. "Den fjärde Reinhardtaftonen. 'Minna von Barnhelm' en stormande succès." *Dagens Nyheter*. 13 November 1915. Svenska Dagstidningar.

Bergman, Gösta M. *Den moderna teaterns genombrott 1890–1925*. Stockholm: Bonniers, 1966.

doi Broomans, Petra, and Marta Ronne. "In the Vanguard of Cultural Transfer." In *In the Vanguard of Cultural Transfer: Cultural Transmitters and Authors in Peripheral Literary Fields*, edited by Petra Broomans & Marta Ronne, 1–12. Groningen: Barkhuis, 2010.

doi Broomans, Petra and Janke Klok. "Thinking about Travelling Ideas on the Waves of Cultural Transfer." In *Travelling Ideas in the Long Nineteenth Century. On the Waves of Cultural Transfer*, edited by Petra Broomans & Janke Klok, 9–28. Groningen: Barkhuis, 2019.

Didring, Ernst. "'Minna von Barnhelm' hos Reinhardt." *Aftonbladet*. 13 November 1915. Svenska Dagstidningar.

doi Drisko, James, and Tina Maschi. *Content Analysis. Pocket Guides to Social Work Research Methods*. New York: Oxford University Press, 2015.

Eisenberg, Christiane. "Kulturtransfer als historischer Prozess. Ein Beitrag zur Komparistik." In *Vergleich und Transfer: Komparatistik in den Sozial-, Geschichts- und Kulturwissenschaften*, edited by Hartmut Kaelble and Jürgen Schriewer, 399–417. Frankfurt: Campus Verlag, 2003.

Ek, Sverker. "Brytningstid." In *1900-talets teater*, edited by Tomas Forser and Sven Åke Heed, 13–34. Ny svensk teaterhistoria. Hedemora: Gidlund, 2007.

Espagne, Michel. "Der Begriff Kulturtransfer." In *Internationale Netzwerke: literarische und ästhetische Transfers im Dreieck Deutschland, Frankreich und Skandinavien zwischen 1870 und 1945*, edited by Karin Hoff, Udo Schöning, & Frédéric Weinmann, 45–56. Würzburg: Königshausen & Neumann, 2016.

Espagne, Michel. "Jenseits der Komparatistik. Zur Methode der Erforschung von Kulturtransfers." In *Europäische Kulturzeitschriften um 1900 als Medien transnationaler und transdisziplinärer Wahrnehmung: Bericht über das Zweite Kolloquium der Kommission 'Europäische Jahrhundertwende – Literatur, Künste, Wissenschaften um 1900 in Grenzüberschreitender Wahrnehmung' (Göttingen, am 4. und 5. Oktober 2004)*, edited by Ulrich Mölk & Susanne Friede, 273: 13–32. Abhandlungen der Akademie der Wissenschaften zu Göttingen, Philologisch-Historische Klasse 3. Göttingen: Vandenhoeck & Ruprecht, 2006.

Espagne, Michel, and Michael Werner. "Deutsch-französischer Kulturtransfer als Forschungsgegenstand: eine Problemskizze." In *Transferts : les rélations interculturelles dans l'espace franco-allemand (XVIIIe et XIXe siècle)*. Paris, 1988.

Fallström, Daniel. "Deutsches Theaters andra föreställning. 'Was ihr wollt' von Shakespeare." *Stockholms-Tidningen*. 11 November 1915. Svenska Dagstidningar.

Fallström, Daniel. "Deutsches Theaters gästspel. Sista föreställningen: Strindbergs 'Oväder'." *Stockholms-Tidningen*. 13 December 1920. Svenskt Teatergalleri. Theaterwissenschaftliche Sammlung Universität zu Köln.

Fallström, Daniel. "Deutsches Theaters gästspel. Strindbergs 'Spöksonaten' andra programmet." *Stockholms-Tidningen*. 4 May 1917. Svenskt Teatergalleri. Theaterwissenschaftliche Sammlung Universität zu Köln.

Fauth, Søren R., Gísli Magnússon, & Peter Wasmus (eds). *Influx: der deutsch-skandinavische Kulturaustausch um 1900*. Würzburg: Königshausen & Neumann, 2014.

Figes, Orlando. *The Europeans. Three lives and the making of a cosmopolitan culture*. New York: Metropolitan Books, 2019.

Fischer-Lichte, Erika. *Ästhetik des Performativen*. 1st ed. Frankfurt am Main: Suhrkamp, 2010.

doi Fischer-Lichte, Erika. "Ibsen's Ghosts – A Play for All Theatre Concepts?: Some Remarks on Its Performance History in Germany." *Ibsen Studies* 7, no. 1 (June 2007): 61–83.

Fischer-Lichte, Erika. "Sinne und Sensationen. Wie Max Reinhardt Theater neu erfand." In *Max Reinhardt und das Deutsche Theater: Texte und Bilder aus Anlass des 100-jährigen Jubiläums seiner Direktion*, edited by Roland Koberg, Bernd Stegemann, & Henrike Reinhardt Thomsen, 13–27. Blätter des Deutschen Theaters 2. Berlin: Henschel, 2005.

doi Greenblatt, Stephen. "A Mobility Studies Manifesto." In *Cultural Mobility*, by Stephen Greenblatt et al., 250–53. Cambridge: Cambridge University Press, 2009.

doi Greenblatt, Stephen. "Cultural Mobility: an introduction." In *Cultural Mobility*, by Stephen Greenblatt et al., 1–23. Cambridge: Cambridge University Press, 2009.

doi Groß, Martina. "Travel Literature and/as Transnational Theatre History – Beyond National Theatre Cultures." In *The Transnational in Literary Studies: Potential and Limitations of a Concept*, edited by Kai Wiegandt, 124–41. Berlin, Boston: De Gruyter, 2020.

Gustafsson Berg, Ruben. "Ibsens 'Gengångare' på Blancheteatern. Moissi som Osvald." *Aftonbladet*. 27 April 1921. Svenskt Teatergalleri. Theaterwissenschaftliche Sammlung Universität zu Köln.

Gustafsson Berg, Ruben. "Reinhardts fredagspremiär. Strindbergs 'Pelikanen' framföres under extatiskt bifall." *Aftonbladet*. 11 December 1920. Svenska Dagstidningar.

doi Hoff, Karin, Anna Sandberg, & Udo Schöning. "Einleitung." In *Literarische Transnationalität: kulturelle Dreiecksbeziehungen zwischen Skandinavien, Deutschland und Frankreich im 19. Jahrhundert*, edited by Karin Hoff, Anna Sandberg, and Udo Schöning, 11–22. Würzburg: Königshausen & Neumann, 2015.

Hoff, Karin, Udo Schöning, & Frédéric Weinmann. "Einleitung/Introduction." In *Internationale Netzwerke: literarische und ästhetische Transfers im Dreieck Deutschland, Frankreich und Skandinavien zwischen 1870 und 1945*, edited by Karin Hoff, Udo Schöning, & Frédéric Weinmann, 7–29. Würzburg: Königshausen & Neumann, 2016.

Jelavich, Peter. "German Culture in the Great War." In *European Culture in the Great War: The Arts, Entertainment, and Propaganda, 1914–1918*, edited by Aviel Roshwald, Richard Stites, & Peter Jelavich, 32–57. Studies in the Social and Cultural History of Modern Warfare 6. Cambridge: Cambridge University Press, 1999.

doi Jiresch, Ester. *Im Netzwerk der Kulturvermittlung: Sechs Autorinnen und ihre Bedeutung für die Verbreitung skandinavischer Literatur und Kultur in West- und Mitteleuropa um 1900*. Groningen: Barkhuis, 2013. http://search.ebscohost.com.proxy-ub.rug.nl/login.aspx?direct=true&db=nlebk&AN=1739063&site=ehost-live&scope=site.

doi Jonas, Michael. *Scandinavia and the Great Powers in the First World War*. London: Bloomsbury Publishing PLC, 2019. http://ebookcentral.proquest.com/lib/rug/detail.action?docID=5633107.

Kaifas. "Reinhardts gästspel börjar 2 maj." *Svenska Dagbladet*. 27 April 1917. Svenska Dagstidningar.

Köppl, Rainer Maria. "Das Tun der Narren. Theaterwissenschaft und Theaterdokumentation." In *Theaterwissenschaft heute: eine Einführung*, edited by Renate Möhrmann & Matthias Müller, 315–69. Berlin: Reimer, 1990.

Kotte, Andreas. *Theatergeschichte: eine Einführung*. Köln: Böhlau, 2013.

Krivanec, Eva. *Kriegsbühnen: Theater im Ersten Weltkrieg; Berlin, Lissabon, Paris und Wien*. 37. Bielefeld: Transcript, 2012.

L-n., S. "Det tyska gästspelet. 'Minna von Barnhelm'." *Aftonbladet*. 8 May 1917. Svenska Dagstidningar.

Lüsebrink, Hans-Jürgen. "Kulturtransfer." In *Interkulturelle Kommunikation: Interaktion, Fremdwahrnehmung, Kulturtransfer*, edited by Hans-Jürgen Lüsebrink, Christiane Solte-Gresser, & Manfred Schmeling. 4th revised and extended edition, 143–88. Stuttgart: J. B. Metzler Verlag, 2016.

Marx, Peter W. "Consuming the Canon: Theatre, Commodification and Social Mobility in Late Nineteenth-Century German Theatre." *Theatre Research International* 31, no. 2 (2006): 129–44.

Marx, Peter W. *Max Reinhardt: vom bürgerlichen Theater zur metropolitanen Kultur*. Tübingen: Francke, 2006.

McConachie, Bruce. "New Media Divide the Theatres of Print Culture, 1870–1930." In *Theatre Histories. An Introduction*, edited by Tobin Nellhaus, Sorgenfrei, Carol Fisher, Tamara Underiner, & Bruce McConachie. London: Routledge, 2016.

Mommsen, Wolfgang J. "Einleitung: Die deutschen kulturellen Eliten im Ersten Weltkrieg." In *Kultur und Krieg: die Rolle der Intellektuellen, Künstler und Schriftsteller im Ersten Weltkrieg*, edited by Wolfgang J. Mommsen, 1–16. Schriften des Historischen Kollegs 34. München: R. Oldenbourg Verlag, 1996.

Opera, Kungliga. "*Vår historia*." Accessed 20 May 2019. https://www.operan.se/om-operan/historik/.

Pratt, Mary Louise. "Arts of the Contact Zone." *Profession*, 1991, 33–40. http://www.jstor.org/stable/25595469.

Reinhardt, Max. *Ich bin nichts als ein Theatermann: Briefe, Reden, Aufsätze, Interviews, Gespräche, Auszüge aus Regiebüchern*. Edited by Hugo Fetting. Berlin: Henschelverlag, 1989.

Rock, Freddie. *Tradition och förnyelse: svensk dramatik och teater från 1914 till 1922*. Stockholm: Akademilitteratur, 1977.

Schriewer, Jürgen. "Problemdimensionen sozialwissenschaftlicher Komparistik." In *Vergleich und Transfer: Komparatistik in den Sozial-, Geschichts- und Kulturwissenschaften*, edited by Hartmut Kaelble & Jürgen Schriewer, 9–52. Frankfurt: Campus Verlag, 2003.

Schuberth, Inger. *Schweden und das Deutsche Reich im Ersten Weltkrieg: die Aktivistenbewegung 1914–1918*. Bonner historische Forschungen 46. Bonn: Röhrscheid, 1981.

Smith, Einar. "Moissi på Blancheteatern." *Svenska Dagbladet*. 27 April 1921. Svenska Dagstidningar.

Sylvester. "Omkring Reinhardt." *Aftonbladet*. 4 November 1915. Svenska Dagstidningar.

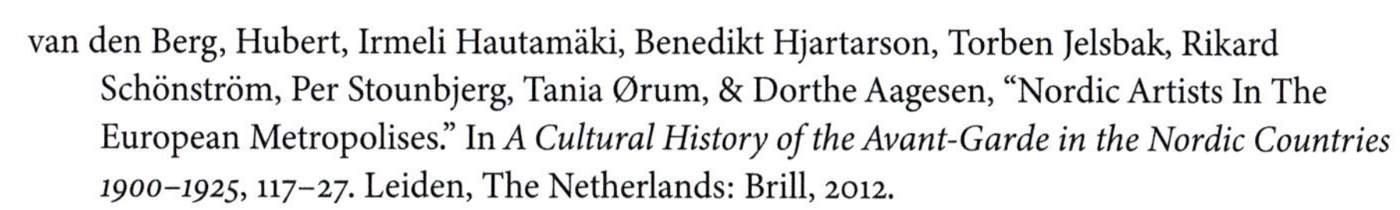

van den Berg, Hubert, Irmeli Hautamäki, Benedikt Hjartarson, Torben Jelsbak, Rikard Schönström, Per Stounbjerg, Tania Ørum, & Dorthe Aagesen, "Nordic Artists In The European Metropolises." In *A Cultural History of the Avant-Garde in the Nordic Countries 1900–1925*, 117–27. Leiden, The Netherlands: Brill, 2012.

van der Poll, Suze, & Rob van der Zalm. "Introduction." In *Reconsidering National Plays in Europe*, edited by Suze van der Poll & Rob van der Zalm, 1–20. Cham: Springer International Publishing, 2018.

Annex I

Productions by Max Reinhardt in Stockholm

Time period	Location	Works	Actors
April 1911	Cirkusteatern Djurgården	*Oedipus Rex*, Sophocles (18.04.)	Company of Deutsches Theater
November 1915	Kungliga Teatern (Operan)	*The Robbers,* Friedrich Schiller (9.11) *Twelfth Night (Was ihr wollt),* William Shakespeare (10.11) *Faust I,* Johann W. von Goethe (11.11, 16.11) *Minna von Barnhelm,* Gotthold Ephraim Lessing (12.11) *A Midsummer Night's Dream,* William Shakespeare (13.11) *The Dance of Death,* August Strindberg (14.11)	Company of Deutsches Theater
May 1917	Kungliga Teatern (Operan)	*Othello,* William Shakespeare (2.5) *The Ghost Sonata,* August Strindberg (3.5) *The Lower Depths (Nachtasyl),* Maxim Gorki (4.5.) *Rose Bernd,* Gerhart Hauptmann (5.5) *Die deutschen Kleinstädter,* August von Kotzebue (6.5.) *Minna von Barnhelm,* Gotthold Ephraim Lessing (7.5.) *The Miracle,* Karl Vollmöller (8.-10.5)	Company of Deutsches Theater
December 1920	Kungliga Teatern (Operan)	*The Merchant of Venice,* William Shakespeare (6.12) *Intrigue and Love,* Friedrich Schiller (7.12) *Er ist an allem Schuld + Die große Szene,* Leo Tolstoy (8.12) *Stella,* J.W. von Goethe (9.12) *The Pelican,* August Strindberg (10.12) *Thunder in the Air,* August Strindberg (12.12)	Company of Deutsches Theater

Annex I *(continued)*

Time period	Location	Works	Actors
October 1921	Kungliga Dramatiska Teatern (Dramaten)	*A Dream Play,* August Strindberg (Premiere: 28.10)	Swedish actors at Dramaten
January 1922	Kungliga Teatern (Operan)	*Orpheus in the Underworld,* Jaques Offenbach (music), Hector Crémieux and Ludovic Halévy (text) (Permiere: 24.01)	Swedish actors and singer at Operan

Performances/productions by Alexander Moissi in Stockholm

Year	Location	Works	Direction	Moissis role
April 1911	Cirkus Orlando	*Oidipus* (18.04)	Max Reinhardt	Oidipus
April 1921	Musikaliska akademien	Reading of several pieces (13.04) Among others: *Es war sein altern König, Mein Kind wir waren Kinder, Die beiden Grenadiere* (Heinrich Heine) *Faust* (Johann W. Goethe) *Der Arbeitsmann* (Richard Dehmel) *Erntelied* (Richard Dehmel) *Novemberwind* (Emilie Verhaerren) *Hobellied* (Ferdinand Raimund) *Schlaflied für Mirjam* (Beer-Hofmann)	–	Reader
April/May 1921	Blancheteatern	*Ghosts,* Henrik Ibsen (Premiere: 26.04) *Er ist an allem Schuld,* Leo Tolstoy (Premiere: 5.05) *Paracelsus,* Arthur Schnitzler (Premiere: 5.05)	Ernst Eklund	Gengångaren: Osvald E.i.a.a.S.: unknown Paracelsus: Paracelsus
October/ November 1921	Blancheteatern	*The Living Corpse,* Leo Tolstoy (Premiere: 13.10)	Alexander Moissi	Fedja
January 1922	Blancheteatern	*Er ist an allem Schuld,* Leo Tolstoy (Premiere: unknown) *Paracelsus,* Arthur Schnitzler (Premiere: unknown)	unknown	

CHAPTER 6

"The East I Know"

Richard Wilhelm and *The Soul of China*

Weishi Yuan
Rheinische Friedrich-Wilhelms-Universität Bonn

This chapter analyses the forms of cultural transfer in Richard Wilhelm's (1873–1930) China travel writings, brought together in *The Soul of China* (1925), with a specific focus on the transmission of the concept of "soul." It examines the two-way transmission of this concept between the East and the West. In *The Soul of China*, Wilhelm created contact zones, which can be understood as a complex entanglement of different mind-sets at the beginning of the twentieth century after a series of social changes in both Germany and China. By means of *The Soul of China*, Wilhelm was determined to appeal to the influential analytical psychology in Europe and to illustrate the ancient spiritual laws of Chinese philosophy, primarily in response to the prevailing European esoteric movement, as well as to the abandonment of Confucianism in China. Wilhelm's earlier translation of *The Secret of the Golden Flower* demonstrated that, in parallel with the soul of the West, there was also a soul of China, which could be understood as consisting of consciousness and unconsciousness. However, part of the soul of China, that element which was shaped by the ancient spiritual laws as found in Wilhelm's translation of the *I Ching*, had not yet been discovered or appropriated by the West. Ultimately, the exchange of the concept of soul was mediated by the construal of a traveller-narrator in the contact zones which Wilhelm created in *The Soul of China*.

Keywords: Richard Wilhelm, the soul of China, *The Secret of the Golden Flower*, *I Ching*, Cultural Transfer, travelogue, *The Soul of China*

https://doi.org/10.1075/fillm.20.06yua

When Richard Wilhelm (1873–1930) published his *Die Seele Chinas* (*The Soul of China*) in German in 1925, "Sinologists," as reported by Wilhelm's wife Salome Wilhelm, "were completely astonished due to the fact that Richard Wilhelm was not just a simple translator and interpreter of Chinese classics; from his mighty and shrewd words, something completely different came off ahead."[1]

For more than three decades, an encounter and exchange with Chinese culture became the focus of Wilhelm's life and writings. In most of his publications, especially his translation of Chinese classics, Wilhelm often translated Chinese knowledge of the spiritual world by referring to Western concepts. However, what Salome Wilhelm suggested in relation to Wilhelm's *The Soul of China* was actually another direction in transculturality that surprised most of the European sinologists. In recounting his many trips to China, rather than merely justifying the concept of soul that had been constructed in a Western context, Wilhelm also intended to enrich this Western concept by referring to the ancient spiritual laws of Chinese philosophy, or more specifically, to classical Confucianism.

Encounter and exchange are particularly present in travel writing, which is an explicitly transcultural genre. Wilhelm's travelogue, *The Soul of China*, was published in 1925 after his return to Germany as a professor of Sinology and was intended to unveil the transcultural exchange of the concept of soul. The greater part of this travelogue consisted of numerous diaries and travel reports of his exploratory journeys over 25 years, in which Wilhelm not only witnessed the collapse of the German colonial system, but also experienced China's transformation from a monarchy to a republic.

The texts in *The Soul of China* took the form of life writing which primarily dealt with the recordings of Wilhelm's eye-witness-accounts: the times in which he lived, the social transformation he experienced, the places he visited and the people he met in China. Wilhelm's travel writing, which were gathered in *The Soul of China*, encompassed the description of a brief history of China as well as of specific events within a time span of over 25 years. At the same time, his travelogue contained his narration of the life of political institutions as well as the lives of individuals in Chinese society. More than just a historical specialism with a chronological overview of events, Wilhelm's life writing also involved his reflections on his encounters with the Chinese people and historical incidents as well

1. Richard Wilhelm, *Die Seele Chinas* (Frankfurt am Main: Insel-Verlag, 1980), 7. (trans. by author). "»Die Sinologen«, so notierte Richard Wilhelms Frau Salome in der Biographie ihres Mannes, »waren erstaunt, daß aus Richards Feder auch einmal etwas ganz anderes hervorging als Übersetzungen der Klassiker und ihre Dichtung. ..." [originally Berlin, 1925; reprint of this edition: Berlin: Reimar Hobbing, 1926]. In this paper, all the citations are from the English translation, *The Soul of China*, trans. John Holroyd-Reece. (New York: Harcourt, Brace and Company, 1928), unless stated otherwise.

as on his psychological and philosophical explorations of the West and the East. The nomadic experience of Wilhelm as a traveller primarily underlined his affinity for Western analytical psychology and ancient Chinese philosophy, especially classical Confucianism. Moreover, his reflections on such nomadic experiences in terms of humanity also led to an attempt to build up reciprocal relationships between cultures, a critical recognition of cultural differences, and broader conceptions of humans.

Primarily known as a translator, throughout his life, Wilhelm made great efforts to introduce Chinese culture to Europeans via his translations and other publications.[2] In addition to translating numerous Chinese classics into German, he also introduced a great amount of Western literature to China. While he originally went to China as a missionary, Wilhelm ultimately became an adherent of Chinese culture. He not only fulfilled his role as a missionary, but was simultaneously translator, cultural advisor, travel writer and university professor. These various roles demonstrate that Wilhelm can be considered a cultural transmitter, usually defined in Cultural Transfer Studies as someone who "works within a particular language and cultural area transmitting one national literature and its cultural context to another."[3]

The fact that Wilhelm adopted the role of traveller-narrator and his cultural translation of ancient Chinese philosophy as carried forwards in *The Soul of China* allowed him to connect European and Chinese perspectives in various ways. Subsequently, in his presentation of ancient Chinese philosophy to the Europeans, the construal of the traveller-narrator mediated the travelling of ideas, or more precisely, the two-way transfer of the concept of soul. On the one hand, Wilhelm justified the existence of the soul of China with analytical psychology which was in favour of Western esoteric tradition. Both in the West and East the soul, which was the essence of the psyche of each individual, was compre-

2. Latest research on Richard Wilhelm's translation can be found in Ruonan Xu, *Klassiker als Brücken zwischen Ost und West: Eine Untersuchung zu Richard Wilhelms Übersetzungsgedanken* (Shanghai: Shanghai Translation Publishing House, 2018); Dorothea Wippermann, *Richard Wilhelm: Der Sinologe und seine Kulturmission in China und Frankfurt* (Frankfurt am Main: Societäts, 2020); Xiaobin Cai, *Richard Wilhelm and Qingdao* (Qingdao: Qingdao Publishing House, 2018); Yuan Tan, "Die Geburt der 'taoistischen Bibel'. Zu Richard Wilhelms Taoteking-Übersetzung," in *Literaturstraße: Chinesisch-deutsches Jahrbuch für Sprache, Literatur und Kultur Band 17•2016*, ed. Feng Yalin, Zhu Jianhua, Wei Yuqing, Gerhard Lauer and Gertrude Roesch (Würzburg: Königshausen & Neumann 2017): 353–367 and Jin'ge Niu, "The Century-Long Dissemination and Impact of Chinese Folk Fairy Tales by Richard Wilhelm," *International Sinology* 40, no.1 (February 2024): 46–54.

3. Petra Broomans, "Introduction," in *From Darwin to Weil. Women as Transmitters of Ideas*, ed. Petra Broomans (Groningen: Barkhuis, 2009), 2.

hended as consisting of consciousness and unconsciousness. On the other hand, he seemed to become a missionary the other way around: he missionised in Germany and tried to "convert" Europeans to the Chinese beliefs – the unique characteristics of the soul of China which were determined by classical Confucianism. Over 25 years, however, he discovered the soul he considered to be specific to China during his travels there. Ultimately, his travelogue allowed the diffusion of this concept of the 'soul of China' within the West. Thereby, Wilhelm also enriched the understanding of the soul in general, which had first been constructed in a Western context. The characteristics of the soul of China – whose consciousness and unconsciousness were particularly shaped by Confucianism – were an asset for Europeans, according to Wilhelm, and should be included in the Western understanding of the soul and of the souls of human beings as such.

Clearly, Wilhelm's unique life experience enabled him to become a mediator between the Western and the Eastern worlds and to transmit one cultural context to another. Moreover, his work as a cultural transmitter usually took shape through contact zones, defined by Mary Louise Pratt as "social spaces where cultures meet, clash, and grapple with each other, often in contexts of highly asymmetrical relations of power."[4] In *The Soul of China*, it is apparent that over the course of time Wilhelm experienced various transformations in both the West and the East. Therefore, the contact zones that were created by Wilhelm in order to transfer the concept of soul between the West and East can be understood as a complex entanglement of different mind-sets, consisting of the intertwinement of multiple struggles in the construction of the self and other.

As the above suggests, the analysis of Wilhelm's travelogue, *The Soul of China*, in this chapter is based on a Cultural Transfer approach which is particularly conceived here in terms of the mobility of a concept, drawing on Stephen Greenblatt's reflection on "cultural mobility." The study of cultural mobility, Greenblatt suggests, should not always concentrate on "theories of hybridity, network theory, and the complex 'flows' of people, goods, money, and information."[5] It must "rethink the fundamental assumption about the fate of culture in an age of global mobility."[6] Hence, this chapter argues that in order to better understand the complexities of cultures and to overcome the challenge of cultural orientation at the age of globalisation, it is necessary to rethink the complex ideological dimensions which facilitate the mobilisation of ideas. In *The Soul of China* Wilhelm's narration of journeys

4. Mary Louise Pratt, "The Art of The Contact Zone," *Profession* (1991): 34.

5. Stephen Greenblatt, "Cultural mobility: an introduction," in *Cultural Mobility: A Manifesto*, eds. Stephen Greenblatt, Ines Županov, Reinhard Meyer-Kalkus, Heike Paul, Pál Nyíri and Frederike Pannewick (Cambridge: Cambridge University Press, 2009), 1.

6. Greenblatt, "Cultural mobility," 2.

and his personal way in which he reported on his encounters in China provide valuable sources to study the travelling of the concept of soul between the West and the East. By revealing how Wilhelm's travelogue reconstructed ancient Chinese spiritual laws in the terms of analytical psychology, a reading of *The Soul of China* contributes to an understanding of the complexities of cultures and cultural transfer.

The first section of this chapter provides a short biography of Wilhelm as a translator and examines the profound influence that his translations had on Western academia. Using two of his other publications, *Das Geheimnis der goldenen Blüte: Das Buch von Bewusstsein und Leben* (*The Secret of the Golden Flower*) and *I Ching: Das Buch der Wandlungen* (the *I Ching*), the second section will focus on analytical psychology and the ancient principles of Chinese philosophy that Wilhelm applied in *The Soul of China* to facilitate the transfer of the concept of soul in both worlds. With the notion of contact zones, the third section will discuss the specific intellectual atmosphere in which Wilhelm lived, and also provide a historical overview of the cultural contact between China and Germany, as well as Europe generally. The final section will provide a brief analysis of Wilhelm's *The Soul of China*. It aims to demonstrate the two-way transfer of the concept of soul in contact zones, as fuelled by the Western esoteric tradition and classical Confucianism.

Richard Wilhelm: Cultural transmitter between the East and the West

Richard Wilhelm was born in Stuttgart in 1873. In 1899, he was sent by the Allgemeiner Evangelisch-protestantischer Missionsverein[7] to Qingdao, a Chinese port city on the western shores of the Yellow Sea, which had just become a German lease area. However, during his work of over 25 years in China, Wilhelm never converted a single Chinese person to Christianity, but ultimately became an advocate of Confucianism[8] and worked as a professor of Sinology at the University of Frankfurt am

7. The Allgemeiner Evangelisch-protestantischer Missionsverein or the General Evangelical Protestant Mission was first established in Weimar in 1884, before it moved its headquarters to Württemberg. From 1893 on, its Swiss branch was directed by Ernst Buß (1843–1928). In 1929, it was renamed as the Ostasienmission (East Asia Mission). During the Second World War, the German and the Swiss branches separated. Today, the Deutsche Ostasienmission (the German East Asia Mission) and the Schweizerische Ostasienmission (the Swiss East Asia Mission) are the successors of the Allgemeiner Evangelisch-protestantischer Missionsverein.

8. Richard Wilhelm's Chinese name 卫礼贤 was chosen from a Confucian classic. As early as 1903, he published an article on Confucius, in which he stated that Confucius was as great a person as Moses in the Bible. See Richard Wilhelm, *Die Stellung des Konfucius unter den Repräsentanten der Menschheit: Vortrag, gehalten in der Deutschen Kolonialgesellschaft Abteilung Tsingtau*, (Tsingtau: Deutsch-Chinesische Verlagsanstalt, 1903).

Main. His first period of acquaintance with Chinese culture in Qingdao spanned the years between 1899 and 1911. In 1910, he published the analects of Confucius in German as *Kungfutse: Gespräche (Lun Yü)*.[9] In 1911, the revolution ended the reign of the last imperial dynasty in China and led to the foundation of a republic. Consequently, numerous ancient scholars, some of whom used to serve the imperial family, fled from Beijing, the turbulent capital, to Qingdao.

Subsequently, Wilhelm had the opportunity to work with some renowned Chinese scholars on his translation of ancient Chinese classics and he also became involved in the attempted restoration of the monarchy. In 1914, after the outbreak of the First World War, Qingdao was occupied by Japan. In 1919, Wilhelm quit his missionary service and in 1922 took up a new position as cultural advisor in the German embassy in Beijing. Between 1922 and 1924, he spent most of his time in Beijing, which enabled him to make contact with scholars from the new era, such as Liang Qichao (1873–1929),[10] Cai Yuanpei (1868–1940)[11] and Hu Shi (1891–1962).[12] In 1924, he published his translation of *I Ching: Das Buch der Wandlungen*[13] and in the same year he left, returning to Germany to take up his position as honorary professor of Sinology at the University of Frankfurt am Main. He also founded the China Institute, where he organised numerous lectures and exhibitions on Chinese culture for the German public.

For Wilhelm, translation served as a means of cultural transfer. Frequently, he endeavoured to search for similarities in the way the Chinese and the West understood the spiritual world and to capture a universal sense of the human being in his translations of the Chinese classics. Consequently, these similarities that were discovered by Wilhelm in the East and the universal or cosmological significance of this for our understanding of human being that he expressed, further inspired and enabled other European scholars, such as Carl Gustav Jung and Hermann Hesse, to consolidate their own theories and to write their own works.

---

9. The English translation was published under the title, *The Analects of Confucius.*

10. Well known as a social and political activist, Liang Qichao (1873–1929) exerted a tremendous influence on the political reformation in modern China. He lived from the last feudalist dynasty to the early Republic of China. He first advocated for a constitutional monarchy in the reformation of 1898, which was defeated. After the revolution of 1911, he served for the government of the Republic of China.

11. Cai Yuanpei (1868–1940) was a Chinese philosopher and politician. He was well known as an influential educational reformer who implemented his own ideology. He served as president of the Peking University and provided Richard Wilhelm with a teaching position there.

12. Hu Shi (1891–1962) was a Chinese literary scholar, philosopher and politician. He was one of the leaders of the New Culture Movement. He was also recognised as a great contributor to Chinese liberalism and the Chinese language reform.

13. Its later English translation by Cary Baynes was entitled, *The I Ching or Book of Changes.*

Richard Wilhelm (1873–1930)

Before he passed away in 1930, Wilhelm translated nearly all of the influential classics of Taoism and Confucianism. In his translation of the Chinese classics, Chinese philosophical concepts were frequently introduced in a language and narrative with which Europeans were familiar. For example, in his translation of the Taoist classic, *Tao Te King*, he referred to expressions from Goethe's *Faust* and the Lutheran Bible. As such, his translations of the Chinese classics offered Europeans an important access point to the spiritual world of the ancient Chinese and had a tremendous influence on the Western world. His translation of the Taoist classics triggered a Taoist fever in the early twentieth century and inspired many scholars in Europe to undertake further studies, which led to the reinterpretation, rethinking and even resignification of Chinese philosophy.[14] It was then that Ger-

14. In the past, the understanding of Chinese philosophy in the West had been limited. In the sixteenth century, Europeans had already learned about Confucianism from the texts sent from

man scholars restarted paying attention to ancient Chinese wisdom again looking to discover an alternative path for Europe, which, after the First World War, remained a damaged continent.[15]

Through Wilhelm, countless German and European scholars had access to the spiritual world of China for the first time. For example, almost completely relying on Wilhelm's translation of Chinese classics, Albert Schweitzer (1875–1965),[16] German philosopher and Nobel Prize winner, gained an in-depth knowledge of China and started to write a book (unfinished) on the history of Chinese philosophy. Carl Gustav Jung (1875–1961), Swiss psychologist and a close friend of Wilhelm, was inspired by the latter's translation of Taoist classics and further deepened his own famous psychological theory of the "collective unconscious" and "synchronicity."[17] Moreover, Hermann Hesse (1877–1962),[18] German poet and novel-

China by the Jesuits. However, it was relatively late that Europeans discovered Taoism, another influential Chinese philosophical tradition alongside Confucianism. The Taoist doctrines and its mystical approach, especially, met the need of Europeans at the turn of the twentieth century. Wilhelm's translations of the Taoist classics made a tremendous contribution to the introduction of Taoism to the West.

15. The term *fin de siècle*, which referred to the end of the nineteenth century in Europe, not only signified a historical moment, but also a particular sensibility in cultural production. For example, the *fin de siècle* was partly characterised by a striving for modernity, in particular a modern way of thinking. The urgent need for a new way of thinking had been clearly proven by the outbreak of the war. Consequently, many European scholars had switched their attentions to the East.

16. In his condolence letter to Wilhelm's wife Salome after Richard Wilhelm's death in 1930, Schweitzer wrote: "The tragic death of your husband struck me deeply. ... And how much I am indebted to him for my scientific research! All my understanding of Chinese thought stems, as a matter of fact, from Dr Wilhelm. From the beginning, I was impressed by the solidity of his work and the wonderful simplicity of his presentation." (trans. by author). "Die Nachricht von dem Tode Ihres verehrten Mannes hat mich tief bewegt ... Und wie viel verdanke ich ihm wissenschaftlich! Alles was ich über chinesisches Denken weiss, geht eigentlich auf ihn zurück. Immer habe ich die Gediegenheit seines Arbeitens und die wundervolle Einfachheit seiner Darstellung bewundert... ." The original letter in German is kept in the archives of Bayerische Akademie der Wissenschaften or the Bavarian Academy of Sciences and Humanities, Munich, Germany. See Condolence letter from Albert Schweitzer to Salome Wilhelm, 1930. ABAdW NL Wilhelm II Nr. 169. Richard Wilhelm Archives, Bayerische Akademie der Wissenschaften, Munich.

17. In his speech, which was delivered as the principal address at a memorial service for Wilhelm held in Munich in May 1930, Jung devoted a long section to the description of the *I Ching* and its relation to the concept of synchronicity. See Carl Gustav Jung, "Richard Wilhelm: In Memoriam," in *The Collected Works of C.G. Jung. Volume 15. The Spirit in Man, Art, and Literature*, ed. Gerhard Adler and Richard Francis Carrington Hull (Princeton: Princeton University Press, 2014), 53–62.

ist, and another loyal friend of Wilhelm, reviewed almost every single publication of Wilhelm's between 1912 and 1932. The Taoist laws of the unity of opposites can be traced in most of Hesse's novels. Last but not least, it was thanks to Wilhelm that Bertolt Brecht (1898–1956),[19] German poet and playwright, composed his own poems with an adaptation of Taoism.

However, Wilhelm's influence did not reside solely among scholars, philosophers and writers of his time. In 1924, after ten years of effort, he published the German translation, *I Ching: Das Buch der Wandlungen*, which contained the ancient Chinese wisdom. It was also regarded as the foundational text of Chinese culture, as well as the shared starting point for both Confucianism and Taoism. In his translation of the *I Ching*, Wilhelm combined the first part of the hexagrams and the second part of the commentaries in order to transform the *I Ching* into a book of wisdom from which all human beings from various times could benefit.[20]

While only around 5,000 copies were sold in the first ten years after its publication, in 1950, the American psychologist, Cary Baynes (1883–1977), translated Wilhelm's *I Ching* into English. From the 1960s onwards, as a consequence of the growing awareness and critique of Western social norms, Baynes' English translation, *The I Ching or Book of Changes*,[21] gained a wider popularity in the English-speaking world and became the most accessible counter-cultural icon. In 1951, with direct reference to the *I Ching*-derived chance operations, American musical theorist, John Cage (1912–1992), composed his ground-breaking work, *The Music of Changes*, aiming to illustrate charts of the various parameters.

Another musician who was inspired by the *I Ching* was Bob Dylan (*1941). In the bootleg version of his song "Idiot Wind," he wrote, "I threw the I-Ching yes-

---

18. For example, in his novel *Siddhartha,* Hesse frequently presented his interpretation of the Taoist philosophical tradition from China.

19. Although he never visited China, Brecht still managed to acquire a lot of knowledge about Chinese culture, mainly by reading Richard Wilhelm's translations. Brecht wrote numerous poems inspired by his understanding of Taoism. See Heinrich Detering, *Bertolt Brecht und Laotse* (Göttingen: Wallstein-Verlag, 2008).

20. Wilhelm's translation was also a typical example of translation as a means of cultural transfer. The original Chinese *I Ching* consisted of two parts: the first part with 64 hexagrams was considered to have been created in the Western Zhou period (1000–750 BC); the second part with 10 commentaries, better known as the *Ten Wings*, was believed to be edited by Confucius. For example, James Legge (1815–1897), one of the most influential sinologists in Scotland in the nineteenth century, translated the hexagrams and the *Ten Wings* separately. For Legge, the *I Ching* was rather a historical book than a book of eternal wisdom. However, Wilhelm was the first person in the West who combined these two parts together in his translation, partly because of the prevailing esoteric movement.

21. Richard Wilhelm, *The I Ching or Book of Changes*, trans. Cary Baynes (New York: The American-Stratford Press, 1950).

terday, it said there'd be some thunder at the well." In a 1965 interview, he stated that "[t]here is a book called the I Ching, I'm not going to push it, I don't want to talk about it, but it is the only thing that is amazingly true, period, not just for me."[22] It was not just Bob Dylan, but also George Harrison (1943–2001) who borrowed from the *I Ching* to challenge conventional Western values. George Harrison indicated that "I wrote 'While my guitar gently weeps' at my mother's house in Warrington. I was thinking about the Chinese I Ching, The Book of Changes. In the West we think of coincidence as being something that just happens – it just happens that I am sitting here and the wind is blowing my hair, and so on. But the Eastern concept is that whatever happens is all meant to be, and that there's no such thing as coincidence – every little item that's going down has a purpose."[23]

Undoubtedly, it was through his translation of the Chinese classics that Wilhelm primarily succeeded in transmitting the knowledge of the Chinese spiritual world to the West. Moreover, his transmission of the fundamental spiritual laws of Chinese philosophy, namely Taoism and Confucianism, which were best illustrated in the *I Ching*, laid down a solid foundation for other textual production in the West. In other words, as a cultural transmitter, Wilhelm enabled many European intellectuals and artists to further produce other thought-provoking masterpieces in various fields.

Chinese philosophy

Before he published *The Soul of China* in 1925, Wilhelm had begun to translate *The Secret of the Golden Flower* in 1922 and had published the translation of the *I Ching* in 1924. A look at these two translations, which he started and completed before *The Soul of China*, provides great insight into the two-way nature of the cultural transfer of the concept of soul in Wilhelm's exploratory travel writing.

In *The Soul of China*, the transfer of the concept of soul is facilitated by both the approach of analytical psychology and the ancient principles of Chinese philosophy. These two pathways of transfer are exemplified in Wilhelm's previous two translation works, respectively. The first transfer of the concept of soul, from the West to the East, was elucidated by Wilhelm in his translation of *The Secret of the Golden Flower*, a book focusing on the esoteric circle, in which Wilhelm recognised the psyche, whose essence later became the soul of China. More precisely, he constructed the "soul" of China using a framework of analytical psy-

22. Jonathan Cott, *Bob Dylan: The Essential Interviews* (New York: Simon & Schuster, 2017), 63.

23. The Beatles, *The Beatles Anthology* (San Francisco: Chronicle Books, 2002), 306.

chology. At the same time, the transfer of the concept of soul from East to West occurred through Wilhelm's interpretation of the spiritual laws of Confucianism in his translation of the *I Ching*. In this translation, he introduced the fundamental principles of Chinese philosophy, or more specifically, classical Confucianism, revealing how they constituted the soul of China, thereby enriching the understanding of the soul by the West. The dimension of the soul that he found in China, according to Wilhelm, could and should also become part of the soul of the West.

The Secret of the Golden Flower

In 1929, working with the Swiss psychologist, Carl Gustav Jung (1875–1961), Wilhelm published *Das Geheimnis der Goldenen Blüte: Das Buch von Bewusstsein und Leben*, which concentrated on an explanation of the Chinese esoteric circle.[24] This publication consisted of Wilhelm's translation of two texts from religious Taoism: *T'ai I Chin Hua Tsung Chin* and *Hui Ming Ching*,[25] and Jung's commentary. Although the publication in German first appeared in 1929, Wilhelm had already started working on the texts – which belonged to the ancient Chinese esoteric circle – in 1922, having received copies of the two original texts on his return to China. After reading Wilhelm's translation in 1928,[26] Jung was inspired to consolidate his theory of the collective unconscious, which he had worked on for years. The *Geheimnis der Goldenen Blüte* was first translated into English by Cary Baynes in 1931 under the title *The Secret of the Golden Flower: A Chinese Book of Life*.

In his translation, in order to demonstrate "the secret of the powers of growth latent in the psyche,"[27] Wilhelm interpreted the Chinese wisdom in the text

---

24. In China, the tradition of Taoism consisted of philosophical Taoism and the later Taoist religion. Philosophical Taoism was considered to be the origin of the Chinese esoteric tradition, with its most famous work, *Tao Te King*. Later on, Taoism was seen as religionised Taoism, which was also under the influence of Buddhism. In the context of religious Taoism, Chinese alchemy consisted of the early branch of *Waidan* (external alchemy or external elixir) and the later branch of *Neidan* (internal alchemy). While *Waidan* was known as a branch of alchemy concentrating on elixirs of immortality created by heating various natural substances, *Neidan* borrowed the vocabulary of *Waidan* and underlined an allegorical elixir within the practitioner's body, via meditation and other physiological practices. Here, the esoteric text of *T'ai I Chin Hua Tsung Chin* that Wilhelm and Jung worked on, belonged to *Neidan*.

25. *T'ai I Chin Hua Tsung Chin* was the text of the school of religious Taoism from the seventeenth century. It taught circulating illumination through the human body until the "golden flower" was formed. Immortality is produced from the "golden flower."

26. Carl Gustav Jung, *The Red Book: Liber Novus*, trans. Sonu Shamdasani (London: W.W. Norton & Company, 2012), 555.

through the lens of psychological experience. Consequently, in Wilhelm's transmission of the Chinese esoteric tradition, the Chinese term "Xing," which was regarded as naming the essence of the psyche, was considered to be composed of consciousness and unconsciousness. In China, it was believed that the cosmos and the human followed common laws, and this cosmological and psychological premise was the Tao. In this understanding, the inner nature of the human was the phenomenal form of the Tao, which developed into a multiplicity of human individuals.[28] In addition, Wilhelm called the monad[29] of each individual the "Lebensprinzip" (life-principle), which separated into "Xing" ("Wesen" or essence) and "Ming" ("Leben" or life) at the moment of conception, thus before birth. Furthermore, the characteristics of essence consisted of those which constituted "Xin" ("Herz" or heart) and those which constituted "Geborenwerden" (origin). The essence of the individual, in Wilhelm's translation, consisted of a conscious part[30] and a substratum[31] which was the superconscious condition (the unconsciousness).

In his later commentary, based on his solidly developed psychological system, Jung improved Wilhelm's first translation by underlining that the psyche consisted of consciousness and unconsciousness. Eventually, this common substratum which constitutes the human being's psyche was called the collective unconscious by Jung,[32] for whom "all conscious imagination and action have grown out of these unconscious prototypes and remain bound up with them."[33] As a cultural transmitter Wilhelm translated the language of Chinese religious Taoism by referring to analytical psychology which prevailed at the beginning of the twentieth century in Europe. In his translation of *The Secret of the Golden Flower* the West and the East shared the same understanding of the psyche, whose essence was

27. Richard Wilhelm and Carl Gustav Jung, *The Secret of the Golden Flower*, trans. Cary F. Baynes (London: Lund Humphries, 1931), viii.

28. Wilhelm and Jung, *The Secret of the Golden Flower*, 13.

29. Wilhelm and Jung, *The Secret of the Golden Flower*, 13.

30. In Wilhelm's translation, essence consisted of heart, which was the seat of emotional consciousness. The English translation of Wilhelm's German translation is: "The heart, according to the Chinese idea, is the seat of emotional consciousness, which is awakened through feeling reactions to impressions received from the external world by the five senses." Wilhelm and Jung, The Secret of the Golden Flower, 13.

31. In Wilhelm's translation, essence also consisted of the superconscious substratum. The English translation of Wilhelm's German translation is: "That what remains as a substratum when no feelings are being expressed, but which lingers, so to speak, in a transcendental, superconscious, condition, is essence." Wilhelm and Jung, The Secret of the Golden Flower, 14.

32. Wilhelm and Jung, *The Secret of the Golden Flower*, 83.

33. Wilhelm and Jung, *The Secret of the Golden Flower*, 84.

later called the soul in *The Soul of China*. In summary, the soul of China delineated the same conceptual dimensions as that of the West.

The *I Ching*

In *The Soul of China*, as well as his translation of the *I Ching*, Wilhelm demonstrated the same structure and the fundamental principles of an ancient Chinese philosophy which had remained unchanged over thousands of years of Chinese culture and history. This ancient philosophy was the origin of both Confucianism and Taoism.[34] According to this philosophy, in general, the ultimate reality is not understood as a quiescent or latent condition, but resides in spiritual law, from which events receive their significance and their impulse towards constant change.[35] Hence, change becomes the central theme of Chinese philosophy. While it is difficult to foresee change, the rules of change can still be learned. Historically, a cosmic foundation with three world forces, heaven, earth and the human, was understood by the ancient Chinese state to have been astrologically conditioned. Heaven creates temporal events and earth is the creative power of expansion, while in Confucianism, the human's function was solely to bring heaven and earth into harmony.[36]

"Never a plain which is not succeeded by a decline, never a journey which is not followed by a return: that is the boundary between Heaven and Earth."[37] This ancient Chinese philosophy portrayed in the *I Ching* was also characterised by a philosophy of nature in which there was a highest impersonal Being called

34. The *I Ching* explained the origin of Chinese philosophy. Primarily, Wilhelm explained the common beliefs, which were shared by both Taoism and Confucianism. However, Wilhelm's translation of the *I Ching* was based on the comments of Confucius. That is to say, apart from the fundamental beliefs – i.e., the theme of change, world affirmation, an optimistic life view and the cosmic foundation of heaven, earth and the human, etc. – the influential spiritual laws that were illustrated in Wilhelm's *I Ching* were mainly shaped by Confucianism. Therefore, the East to West transfer of the concept of the soul (of China), or the transmission of the ancient principles of Chinese philosophy in *The Soul of China*, merely concerned the transmission of the spiritual laws of classical Confucianism to the West.

35. Hellmut Wilhelm, "Change: Eight Lectures on the I Ching," in *Understanding the I Ching. The Wilhelm Lectures on the Book of Changes*, eds. Helmut Wilhelm and Richard Wilhelm (Princeton: Princeton University Press, 1995), 20–32.

36. As suggested in a footnote above, although Wilhelm agreed with most of the Chinese scholars that *I Ching* contained the basic spiritual laws from which both Confucianism and Taoism were derived, he solely referred to the exegesis, which was the work of Confucianism, namely of Confucius. That is to say, in his translation of the *I Ching*, Wilhelm obviously placed greater weight on the spiritual laws of classical Confucianism.

37. Richard Wilhelm, *The Soul of China*, trans. John Holroyd-Reece (New York: Harcourt, Brace and Company, 1928), 357.

heaven. In Chinese philosophy, the natural order of the universe was the Tao,[38] which was interpreted by Taoism as "Sinn" (meaning),[39] but by Confucianism as the "göttliche Vernunft" (divine rationality).[40] The Tao, which contained two forces of Yin and Yang, was encapsulated in heaven. In Taoism, this heaven was frequently identified as the physical heavens, while Confucianism was more inclined to understand heaven, which is often equated with the concept of God in the West, as a sort of sum total of virtues. "Humanity's natural endowment is defined by Heaven. To act according to his nature is called the Tao (of heaven), which serves as the means to regulate natural events and Human life. To develop and to attain the Tao is dependent on moral education."[41] In particular, with regard to the origin of virtues, it is said that "God (heaven) begot the (ethical) spirit in me."[42] According to Confucianism, heaven engendered virtues in the human being and the real significance of human virtues was to follow the example of the Tao of heaven.[43] Furthermore, the genuine social function of the human virtues was that: "He [the human] educates himself in moral earnestness...in order to bring peace to the hundreds of others."[44] It is solely through actions which assisted human generation and brought comfort to other human beings that an individual would be able to arrive at an ethical personality. Accordingly, the best way to govern a country was by the virtue of "one's own nature,"[45] not

38. Over the course of time, Tao was translated into various different European words, for example, *"Gott"* (God), "Weg" (route), "Vernunft" (rationality) or "Wort" (word).

39. Richard Wilhelm, *Tao Te King: Das Buch des Alten vom Sinn und Leben* (Jena: Eugen Diederichs, 1911), XV.

40. Richard Wilhelm, *Kungfutse: Gespräche (Lun Yü)* (Jena: Eugen Diederichs, 1910), X.

41. Richard Wilhelm, *Li Gi: Das Buch der Sitte des älteren und jüngeren Dai* (Jena: Eugen Diederichs. 1930), 3. (trans. by author). "Was der Himmel (dem Menschen) bestimmt hat, ist sein Wesen. Was dieses Wesen (zum Rechten) leitet, ist der Weg. Was den Weg ausbildet, ist die Erziehung."

42. Richard Wilhelm, *Kungfutse: Gespräche (Lun Yü)* (Jena: Eugen Diederichs, 1910), 69. (trans. by author). "Der Meister sprach: 'Gott hat den Geist in mir gezeugt....'"

43. Richard Wilhelm, *The I Ching or Book of Changes*, trans. Cary Baynes (New York: The American-Stratford Press: 1950), 283. This chapter of *I Ching* illustrated the fundamental principles regarding the Tao: "The Tao of Heaven was the dark and the light; the Tao of the Earth was the yielding and the firm; the Tao of Man was love and rectitude. The *I Ching* made these three fundamental powers and doubled them; therefore, in the *Book of Changes* a sign was always six lines. As the Tao of Heaven was the dark (Yin) and the light (Yang), in the Tao of the Earth, the yielding corresponded to the dark while the firm corresponded to the light. Accordingly, in the Tao of Man, love corresponded to the dark and rectitude corresponded to the light."

44. Richard Wilhelm, *Kungfutse: Gespräche (Lun Yü)* (Jena: Eugen Diederichs, 1910), 167. (trans. by author). "Er bildet sich selbst aus (sittlichem) Ernst. Er bildet sich selbst, um andern Frieden zu geben. Er bildet sich selbst, um den hundert Namen Frieden zu geben."

legal institutions. Certainly, "the people could attain a conscience (of being ethical) and eventually strive for the good, if a ruler runs a country simply by virtue of his natural endowment and organises the society according to moral standards and customs."[46] An improvement in moral accomplishment was mostly dependent on a human's own moral education. As a result, Confucianism attached great importance to moral education and self-cultivation. It also emphasised the role of government, which was particularly dominated by the ruler's moral standards. In a cosmic dimension, a human was never considered in isolation from other humans but always in connection with many.

In this regard, Taoism can therefore be classified as a form of mysticism,[47] in the sense of it aiming for the union with what has been variously named the Absolute, the Infinite, or God. By contrast, Confucianism, which can be classified as a form of rationalism, believed that the realisation of the union with heaven or the Tao of heaven could only succeed through the practice of the Tao by humans. What was deemed to constantly change was the means which enabled the ultimate realisation of the ideal moral standards. However, the basic moral principles should not alter over the course of time.

Whether in the form of Taoist mysticism or that of Confucian rationalism, humankind will find "a harmonious existence by virtue of a profound acceptance and affirmation of the world as his cosmos."[48] According to Confucianism, harmony is the ultimate force, which can only exist through the constant transcendence of the nature of the significance of the world. Meanwhile, the realisation of harmony will usually lead to either motion or transition.[49] Times of peace are

---

45. Wilhelm, *Kungfutse*, 8. "Wer kraft seines Wesens herrscht, gleicht dem Nordstern."

46. Wilhelm, *Kungfutse*, 8. "Wenn man durch Erlasse leitet und durch Strafe ordnet, so weicht das Volk aus und hat kein Gewissen. Wenn man durch Kraft des Wesens leitet und durch Sitte ordnet, so hat das Volk Gewissen und erreicht (das Gute)."

47. In contrast to Confucianism, however, Taoism was against any form of moral education and believed that any ethics which was imposed through external influence, was hopeless. It was the conviction of Taoism that the fundamental ethical principle was to act according to the Tao (of heaven). All knowledge, which was derived from "Erkennen" (cognition), must be condemned as insufficient. The ideal ethics, which was suggested by Taoism could only be achieved by "Schauen" (observation) and it must consequently arrive through inward enlightenment. Taoism never preached a logical exploration of natural events. According to Taoism, it is the ultimate destiny of the human to comprehend the Tao of heaven, and its highest form was the union with heaven.

48. Richard Wilhelm, *The Soul of China*, trans. John Holroyd-Reece (New York: Harcourt, Brace and Company, 1928), 358.

49. While Confucianism focuses on the organisation of humanity, Taoism emphasises the general relation of the human to nature. Taoism does not favour the suppression of nature. Perfection depends on human's unification with cosmic forces.

usually followed by a period of chaos, because humans cannot always succeed in achieving harmony. It is only when such changes can be guided in the right way that growth can be expected.

This was the basis of the two-way transfer of the concept of soul in *The Soul of China*: first, the transfer of the concept of soul from the West to the East was undertaken in his translation of *The Secret of the Golden Flower*, a group of texts of ancient religious Taoism, in which Wilhelm interpreted the soul of China through the framework of analytical psychology. Second, the transfer of the concept of soul from the East to the West in his translation of the *I Ching*, in which Wilhelm intended to introduce the soul of China as shaped by the spiritual laws of classical Confucianism. This two-way transfer of the concept of soul, as represented in these two translations, ultimately took shape through contact zones which were predominantly influenced by the esoteric movement and classical Confucianism, leading to the new work *The Soul of China*.

Contact zones: Cultural contact between China and Germany/Europe

The Soul of China consists of Wilhelm's reports of his exploratory journeys in China between 1899 and 1924. Over a period of 25 years, Wilhelm witnessed the collapse of the German colonial system and the rise of the esoteric movement, which intended to break the hegemony of the rational pursuit of knowledge. Various new approaches had been adopted by Western scholars searching for knowledge with regard to the nature of the world and the role of humanity. At the same time, Wilhelm also experienced China's transformation from a feudal state to a republic. In this context, the authority of Confucianism was severely challenged. Moreover, the time in which Wilhelm lived was still characterised by a Western hostility towards China, although there was already a shift of opinion away from Sinophobia towards Sinophilia.

Thus, during Wilhelm's time in China, there was not only a clash between the past and the present in both Europe and in China, but also a struggle to appreciate Chinese culture under the prevailing influence of Sinophobia in Europe. Having embraced a Western esoteric approach in his collaboration with Jung, Wilhelm then applied this approach to his own understanding of the concept of soul in the Chinese context, as discussed in the previous section. In this interpretation of the soul of China, Wilhelm placed great value on Confucianism, whose doctrines, he recognised, had played a significant role in shaping the soul of China and should also be learned by the West. Last but not least, by revealing and advocating the soul of China, he recognised that he could enrich the Western concept of soul, thus suggesting that Chinese culture could undoubtedly help the West to improve its self-understanding as well as that of all humankind.

Contact zones: Esotericism and Confucianism

As we have seen, in his presentation of China, Wilhelm used the framework of Jung's analytical psychology, which echoed the esoteric[50] tradition that was prevalent in Europe after the First World War. The term "Western Esotericism" is used to categorise a large number of loosely related ideas or movements which had developed in Western society and that pervaded various forms of religion, philosophy, literature and art. Although the various esoteric schools have different definitions and focuses, they still share certain similar characteristics. The term "esoteric" is of Greek origin and primarily means "belonging to an inner circle." In general, Western esotericism was "concerned with asking and answering questions about the nature of the world, its relation to the divine and the role of humanity in between."[51] In addition, the primary motivation was somewhat religious "in the sense of a deep concern with the true meaning of life and the ultimate spiritual destiny of human beings in the universe."[52]

An esoteric approach could exclusively focus on the interior practices of direct experiences. Furthermore, the ideas or currents which are embraced by esotericism are usually distinct from both the mysticism of orthodox Christianity and the rationalism of the Enlightenment. For example, the publishing house, Eugen Diederichs Verlag, which published most of Wilhelm's translations, was well known for its focus on esoteric literature. In 1920, in the hope of offering a new philosophical method for both teaching and learning, Hermann Keyserling (1880–1946) founded the Schule der Weisheit (School of Wisdom). However, Keyserling said that a governing idea such as wisdom can neither be acquired through systematic intellectual learning nor transmitted by a special capability or isolated knowledge.[53] Over a long time, Wilhelm was frequently invited to give lectures in the Schule der Weisheit. It was Wilhelm's conviction in *The Soul of China* that Chinese philosophy had proved that if humanity desired a release from the ties of time and space, there was a need for "a profound penetration into subconsciousness"[54] in a search for the path to all living beings, which could be "experienced

50. Today, it is believed that it was Richard Wilhelm who first introduced this esoteric tradition from China – for example, the mysticism of Taoism and the rationalism of Confucianism – before it was further theorised by other European scholars, especially by Carl Gustav Jung.

51. Wouter J. Hanegraaff, *Western Esotericism: A Guide for the Perplexed* (London: Bloomsburg, 2013), 69.

52. Hanegraaff, *Western Esotericism*, 69.

53. Cf. Institut für Praxis der Philosophie, "Die Schule der Weisheit," last modified 2020. (trans. by author).

54. Wilhelm, *The Soul of China*, 373.

intuitively in a mystic and unified vision."[55] In summary, the interpretation of the Chinese spiritual world by Wilhelm had a mystical perspective. In other words, the contact zone here was not physically but mystically.

In 1911, the Revolution ended the last feudal dynasty and China became a republic. Confucianism, which provided the belief system for the feudal dynasties and the foundation for much of Chinese culture, was questioned and rejected. Wilhelm had become acquainted with Confucianism when he first arrived in China and, accordingly, he started his translations of the Confucian classics. However, after 1911, he began translating the Taoist classics. Several years later, he also became involved in the restoration of the monarchy in China and returned to the translation of Confucian classics.

The Soul of China, in which Wilhelm transmitted knowledge of ancient Chinese culture to Europe, particularly the philosophy of Confucianism, was written at a time when the whole of Chinese society was gradually becoming unfriendly and even hostile to the existence of Confucianism. In China, especially after 1919, Confucianism had lost its status as the state religion, and it was abandoned by many who wished to cut all ties with the feudal past. At the same time, European intellectuals were discovering the timeless wisdom and the mystical approach of Taoism. However, in *The Soul of China*, it appears that Wilhelm was determined to revive classical Confucianism, despite it being ignored by mainstream academia in both the West and East at this time. In the West, this lack of interest was mainly because China was still considered to be morally inferior and new modes of thought had been found wanting in the East. In this context, Wilhelm's intention to introduce the ancient principles of classical Confucianism to the West suggests that it could also be considered as a contact zone.

Sino-European cultural contact

In the colonial epoch, Jesuit missionaries set out for China in the sixteenth and seventeenth centuries and sent many reports back to Europe about China. Although this was probably the first time that Europe had such thorough and in-depth accounts about China, these reports were primarily produced in the light of their missionary objective. Nevertheless, they also demonstrated great interest in the Chinese spiritual world. For example, the Jesuit missionary, Matteo Ricci (1552–1610), made great efforts to understand the spiritual foundation of Chinese philosophy for the purpose of Christian missionary work. It is fascinating to see to what extent European philosophers from the age of the Enlightenment were

55. Wilhelm, *The Soul of China*, 373.

influenced by the Jesuits' reports on China. In the eyes of Leibniz and Voltaire, China served as a positive example from the non-European, non-Christian world in supporting their criticism of the doctrines of the Church. In the example of China, they saw that it was possible to attain the highest morality without the biblical tradition or even revelation.

The contact between Germany and China in the colonial epoch started before German unification in 1871. Between 1868 and 1872, the German geologist Ferdinand von Richthofen (1833–1905)[56] had conducted seven trips to China, primarily in search of mineral resources. Numerous scientific reports by Richthofen directly served Germany's colonial interests, and his advice to occupy Qingdao was also taken by the German government. Richthofen was also a modern pioneer in descriptive mapping of place, as well as standardising the study of causal relations between geographical phenomena occurring within a particular region. He is often credited with coining the term "Seidenstraße" (the Silk Road), which he first mentioned in his inaugural lecture at the University of Bonn in 1877.[57] Richthofen used the concept to refer to the series of trading routes he had uncovered, another example of contact zones between Europe and China which later gained widespread acceptance in academia and popularity among the public.

Over the next 20 years, due to Germany's expanding colonial interests and ambitions, China was always depicted as a morally inferior other. In 1895, based on a sketch by the German emperor, Wilhelm II, historian Hermann Knackfuß (1848–1915)[58] produced a painting entitled "Völker Europas, wahrt eure heiligsten Güter,"[59] which triggered propaganda about the "Yellow Peril" in the West. Following Germany's defeat in the First World War, its colonial system collapsed, with Germany losing all of its overseas colonies, including the lease area in Qingdao. Moreover, after the catastrophe of the First World War for Europe more

56. Ferdinand von Richthofen (1833–1905) was a German traveller, geologist and scientist. He studied at the University of Breslau and the University of Berlin. Richthofen was well known for his trips around the world in combination with the discovery of minerals. From 1862 to 1868, he worked in the United States and discovered the goldfields in California. His research findings had stimulated the colonial interest of the German government. From 1875 on, Richthofen was appointed professor of geology at several German universities.

57. Matthias Mertens, "Did Richthofen really coin 'the Silk Road'?" *The Silk Road* 17, (2019): 1–9. In China, there was not a specific term for these trading routes. Until recently, it was generally believed that Richthofen first coined the term "Silk Road". However, although Richthofen had tremendous influence in conceptualising and popularising the idea, recent discoveries have indicated that other historians had already used the term in their publications.

58. Hermann Knackfuß (1848–1915) was a German painter and writer. He was well known for his historical paintings.

59. The English translation would be "Peoples of Europe, Guard your Dearest Goods."

broadly, there was a prevailing sense of pessimism among individuals and society, which led to a turn from Sinophobia to Sinophilia in certain fields of literature. However, mainstream academia in Europe was still dominated by the influence of the concept of the Yellow Peril. By contrast, in *The Soul of China*, Richard Wilhelm provided a positive illustration of the Chinese inner spiritual world by reviving China's past.

Evidently, in *The Soul of China*, the contact zone in which the exchange of the concept of soul took place was the site of a struggle for both the construction of the self and the construction of the other. At first, the construction of the West was facilitated by his embrace of the new approach of analytical psychology and his recognition of the profoundness of Chinese culture in assisting the West with its self-understanding. At the same time, the positive[60] construction of the East occurred in the exploration of esotericism and an emphasis on the remarkable role of Confucianism in constantly constituting the fundamental principles of Chinese culture.

The soul of China

The role of Wilhelm as a cultural transmitter is clear in *The Soul of China*, with his reflections on his encounters with the Chinese people and historical incidents in China constantly echoing his interest in the approach of analytical psychology – which was connected with the Western esoteric tradition – and his interpretation of the spiritual laws of ancient Chinese philosophy, especially classical Confucianism. In Wilhelm's travelogue, this aspect of transmission, which was

60. For centuries European portrayal of China alternated with high praise or harsh critiques. Accordingly, China was perceived to be either morally superior or inferior. For example, when Europe was in search of alternative ideas for its own culture which was in crisis, China was usually regarded as land of inspirations. This trend was also recorded by Karl-Heinz Pohl in his "Unser Chinabild – Von Marco Polo bis heute," which provided a thorough analysis of European perception of China from antiquity to the 1950s. See Karl-Heinz Pohl, "Unser Chinabild – Von Marco Polo bis heute," in *China: Gesellschaft und Wirtschaft im Umbruch*, ed. Karin Aschenbrücker and Hansjörg Bisle-Müller (Augsburg: Wißner-Verlag, 2009), 209–220. Certainly, Wilhelm's presentation of Chinese philosophy, especially with the approach of analytical psychology, was also in resonance with the shift of opinion towards Sinophilia at the beginning of the twentieth century in Europe. Although Wilhelm's research, particularly his idealisation of Chinese philosophy received massive criticism from not only his fellow sinologist colleagues, for example, Alfred Forke (1867–1944), Otto Franke (1863–1946) and Erich Hauer (1878–1936), but also contemporary scholars from Europe and China, a scholarship which strives for systematic and critical study of Wilhelm's positive portrayal of Chinese philosophy has not existed yet.

best exemplified in his reflections on his encounters,[61] was mediated by his narration of his journeys in China. This narration became the site of exchange of the concept of soul between the West and the East, and one example – among multiple attempts – of the spiritual (re)construction of the West and the East at the time. For Wilhelm, like the soul of the West, the soul of China was also composed of consciousness and unconsciousness. In the soul of China, he found a unique dimension of this dual consciousness and unconsciousness that was particularly shaped by classical Confucianism, and which he thought should also be adopted in the understanding of the soul by the West.

As suggested above, Wilhelm wrote *The Soul of China* in the aftermath of turbulent transformations in China. However, drawing on the wisdom of Confucianism, Wilhelm also indicated that despite constant changes, "in the old and in the new there was, nevertheless, a common element: the Soul of China in the course of evolution; that soul which had not lost its gentleness nor its calm, and will, I hope, never lose them."[62] In his book, it was his primary aim to introduce China to the Europeans through his interpretation of the soul of China.

When Wilhelm first arrived in China, one of his deepest insights came from observing day labourers (coolies) in the Chinese cities, who were despised by most other Chinese and labelled lazy, impertinent and deceitful. They were generally dealt with through violence. However, in Wilhelm's eyes, "they were all human beings, men with their joys and sufferings, men who had to fight the battle of life, who had to make their way by cunning and patience, and who had to travel their road by straight or crooked paths."[63]

According to Jung's theory, part of human unconsciousness was a sense of inferiority, which was due to having to give up one's dignity in order to survive. However, this inferiority is not easily accepted by human beings. Consequently, in rejecting such inferiority in unconsciousness, humankind consciously appears to be arrogant. That is to say, in Wilhelm's description, the arrogance of those superior Chinese towards the coolies was the expression of their own sense of inferiority which was deeply rooted in their unconsciousness.

Clearly, in his description, Wilhelm suggests that there is definitely a soul of China, when he claims that the Chinese and the Europeans are all human beings. Furthermore, like the soul of the West, this soul of China also consists of uncon-

61. The best examples are: Wilhelm saw and reflected upon his encounter with the Chinese day labourers in Chapter 1, "My Arrival in the East"; and his narration and reflection on his encounter with an old Chinese woman during his visit to the Tai Mountain in Chapter 8, "The Holy Mountain."

62. Wilhelm, *The Soul of China*, 7.

63. Wilhelm, *The Soul of China*, 23.

sciousness and consciousness, with the latter developing from the former. Ultimately, the concept of soul, which was initially used by Wilhelm in the sense of analytical psychology that was favoured by the Western esoteric tradition, was transferred into Chinese culture.

Additionally, Wilhelm's *The Soul of China* included an awareness of an ancient spiritual law, with the main theme of change and harmony, and also a consciousness of a cognitive orientation of individuals and society; for example, how harmony can be attained and how individuals should act in order to achieve it. In the first six chapters of his book, Wilhelm illustrated the turbulent history of China starting with the Opium Wars in the nineteenth century, characterised by conflicts between the monarchy and the Chinese people as well as conflicts between colonial powers and China as a whole. However, there were also times of temporary peace or progress, which occurred either through the restoration of the tradition or the Chinese adaptation of European ideas.

According to Wilhelm's description of the Chinese politicians who exerted instrumental impact on China's history at the turn of the twentieth century, there were constant struggles between them to find a compromise between the political traditions of the past and the desires of the present. Some politicians remained conservative, defended the ancient traditions and saw Western ideas exclusively as bringing undesirable results to Chinese society. At the same time, others strove to attain an ideal system which could retain Confucian moral at its centre and European power at its periphery. Through his travels in time, Wilhelm suggested that, in the soul of China, a consciousness of the guiding forces which served as the means to realise harmony and prosperity varied from time to time.

In *The Soul of China*, Wilhelm also recorded an encounter with an old Chinese woman, whose cheerful face impressed him on a trip to Tai Mountain, a holy mountain in northern China. The pilgrimage upwards was difficult, and the tiny old Chinese woman climbed bravely with her small bound feet. When Wilhelm inquired about the reason for such a pilgrimage, she replied: "It is not hard, the pilgrimage; I am now seventy years old and my life lies behind me. There is nothing I have to hope for or fear; but that I have got so far, that the old Lord of Heaven has helped me through all these many years when I brought up my sons and grandchildren and through all the misery and burden of life, makes me grateful. And at the end, I want to show this to him. I ask nothing and I avoid nothing. I am quite quiet, and so the journey was not hard."[64]

Like the story of this old Chinese woman, there are many examples of this virtue of acceptance in Wilhelm's narration of his journeys. He sees in them the affirmation of the world as their cosmos and considers their life view to be opti-

64. Wilhelm, *The Soul of China*, 124.

mistic because they believe that in the long run, as the *I Ching* taught, everything would be again brought back into harmony. All of these stories depict what Wilhelm called the gentleness and calm which had never been lost by the soul of China. In the narration of his travels over 25 years and his encounter with the old woman, Wilhelm suggests there is a consciousness of guiding forces allowing one to attain prosperity, an unconscious affirmation of the cosmos as humanity's own cosmos and the affirmation of the world in humanity's optimistic life view. This understanding of the soul of China, which was composed of consciousness and unconsciousness shaped by Confucianism, should and could, according to Wilhelm, also be learned by and discovered in the soul of the West.

The region of Qingdao, where Wilhelm spent most of his time in China, was located in the same province in which Confucius was born. After his visit to the birthplace of Confucius, Wilhelm gave full expression to his opinion on the role of Confucianism in the new era after the revolution in 1911. During his sojourn at a temple dedicated to Confucius, Wilhelm's attention was caught by the ancient cypress trees which had kept guard over this temple for centuries. While the dry trunks of some of the trees were crumbling and collapsing, the roots of others had produced new shoots. Indeed, in China at the time, the spiritual world that had been based on and shaped by Confucianism had now started to collapse and had to give way to new trends of thought. However, beyond this pessimistic complaint about the defeat of the glorious past, Wilhelm also discovered a historical resemblance between the ancient times, in which Confucius lived, and the violent transition from a monarchy to a republic, which Wilhelm experienced. In a turbulent period of the past, in which the structure of contemporary culture had begun to be shaken to its foundations, Confucius provided a plan, according to which – "subject to different circumstances and yet consistent with the spirit of his heritage"[65] – the culture of China could be revived once again. Although it seemed that there was little or no hope that Confucius' preaching would not be dissolved in this new era, it was Wilhelm's belief that the great truths of Confucianism, especially the truth of the harmony between nature and culture, would remain eternally into the future. Certainly, for him, the truth of Confucianism would always provide new impulses for the new philosophy as well as for the future development of human beings.

65. Wilhelm, *The Soul of China*, 108.

Conclusion

In response to a series of social transformations in Europe and China, Wilhelm embraced the prevailing western esoteric tradition and defended Confucianism. And this stance led to his focus on Jungian analytical psychology and the spiritual laws of ancient Chinese philosophy. While he mediated between analytical psychology and ancient Chinese philosophy of Confucianism, Wilhelm interpreted the Chinese inner spiritual world from a mystical perspective.

By referring to the texts of religious Taoism, Wilhelm's cultural translation of the form of ancient Chinese philosophy into consciousness and unconsciousness eventually arrived at the concept of the soul of China. This was significantly viewed through the lens of the Jungian approach of analytical psychology and the esoteric movement in Europe. At the same time, Wilhelm became involved in the attempted restoration of the monarchy in China, in accordance with his desire to save the treasures of Confucianism, which was at that time thought by many to be in great danger.

In *The Soul of China*, he intended to make the Europeans familiar with the ancient Chinese spiritual laws, with their main theme of change, world affirmation and optimistic life view, which were shared by both Taoism and Confucianism in the light of the *I Ching*. He further defended the values of Confucianism, the theme of harmony and a consciousness of the cognitive orientation of individuals and society. However, he also came to the conclusion that, while a period of irreversible change had arrived, the fundamental teachings of Confucianism could still exert a positive influence in the future by contributing to the formation of a new philosophy. With the experience he had gained, and a sense of the time span from the old to the modern era, Wilhelm was able to look beyond the surface and to discover the profundity of Chinese culture.

The contact zones, which were created by Wilhelm in his travelogue to facilitate the mobilisation of the concept of soul, were not physical. Rather, it was a complex entanglement of different mind-sets and became the place where both western and Chinese values met, confronted and interacted with each other. Consequently, such a cultural transfer between the West and the East was crystalised by the example of the two-way exchange of the concept of soul.

In *The Soul of China*, the transfer of the concept of soul was mediated by the construal of the traveller-narrator. Wilhelm's mediation between analytical psychology and the ancient Chinese philosophy of Confucianism resulted in a combination of both the perceptions of analytical psychology and the spiritual laws of Confucianism to analyse his encounters with the Chinese people and his experience of historical incidents. In his narrations, there was clearly an attempt to build up reciprocal relationships between the West and the East: the West and the East were indispensable to each other. Both the West and the East could provide either a reference or a supplement to each other in order to better understand themselves. The different spiritual laws of ancient Chinese philosophy became an asset for the West. A new approach of western analytical psychology in order to capture the real meaning of life and understand the spiritual destiny of human beings was justified and further empowered by the Chinese wisdom. Both western psychology and ancient Chinese philosophy could equally contribute to the development of human kinds.

At first this preference of analytical psychology was to emphasize the interior practices of direct experiences and to justify the existence of the soul of China. The soul of China, which Wilhelm understood to consist of consciousness and unconsciousness, was first transmitted to the West by Wilhelm. He believed that the soul of China possessed the same dimension as that of the West. However, the soul of China also displayed certain unique characteristics in its understanding of consciousness and unconsciousness.

“In China people count by centuries.”[66] This statement by ancient colonists, quoted by Wilhelm, showed their conviction that Chinese culture seemed destined to last for centuries upon centuries. Yet, after the revolution, as Wilhelm noted, life in China, which had seemed so constant and steady, now “proceeds at a feverish speed.”[67] In Wilhelm's time, the soul of China was in constant transformation, with the consciousness of the guiding forces among individuals and society – forces which eventually bring prosperity – changing from time to time. The guiding forces were best known as the Tao of the human in the *I Ching* and explained by Confucianism as corresponding to ethical actions after rational contemplation. However, the soul of China also contained certain unchanged elements – the unconscious affirmation of the cosmos as well as the affirmation of the world as the human cosmos and an optimistic life view. This understanding of changing consciousness and unchanged unconsciousness in the soul of China, which was shaped by the ancient Chinese philosophy of Confucianism, should, according to Wilhelm, also be learned and explored by the West.

References

Broomans, Petra. “Introduction.” In *From Darwin to Weil. Women as Transmitters of Ideas*, edited by Petra Broomans, 1–20. Groningen: Barkhuis, 2009.

Cott, Jonathan. *Bob Dylan: The Essential Interviews*. New York: Simon & Schuster, 2017.

66. Wilhelm, *The Soul of China*, 7.
67. Wilhelm, *The Soul of China*, 7.

Detering, Heinrich. *Bertolt Brecht und Laotse*. Göttingen: Wallstein-Verlag, 2008.

Greenblatt, Stephen. "Cultural mobility: an introduction." In *Cultural Mobility: A Manifesto*, edited by Stephen Greenblatt, Ines Županov, Reinhard Meyer-Kalkus, Heike Paul, Pál Nyíri and Frederike Pannewick, 1–23. Cambridge: Cambridge University Press, 2009.

Hanegraaff, Wouter J. *Western Esotericism: A Guide for the Perplexed*. London: Bloomsburg, 2013.

Institut für Praxis der Philosophie. "*Die Schule der Weisheit*", last modified 2020. http://www.ipph-darmstadt.de/schule-der-weisheit/

Jung, Carl Gustav. "Richard Wilhelm: In Memoriam." In *The Collected Works of C.G. Jung. Volume 15. The Spirit in Man, Art, and Literature*, edited by Gerhard Adler and Richard Francis Carrington Hull, 53–62. Princeton: Princeton University Press, 2014.

Jung, Carl Gustav. *The Red Book: Liber Novus*. Translated by Sonu Shamdasani. London: W.W. Norton & Company, 2012.

Mertens, Matthias. "Did Richthofen really coin 'the Silk Road'?" *The Silk Road* 17, (2019): 1–9.

Pohl, Karl Heiz. "Unser Chinabild – Von Marco Polo bis heute." In *China: Gesellschaft und Wirtschaft im Umbruch*, edited by Karin Aschenbrücker and Hansjörg Bisle-Müller, 209–220. Augsburg: Wißner-Verlag, 2009.

Pratt, Mary Louise. "The Art of The Contact Zone," *Profession* (1991): 34.

Schweitzer, Albert. *Condolence Letter from Albert Schweitzer to Salome Wilhelm*, 1930. ABAdW NL Wilhelm II Nr. 169. Richard Wilhelm Archives, Bayerische Akademie der Wissenschaften, Munich.

The Beatles. *The Beatles Anthology*. San Francisco: Chronicle Books, 2002.

Wilhelm, Hellmut. "Change: Eight Lectures on the I Ching." In *Understanding the I Ching. The Wilhelm Lectures on the Book of Changes*, edited by Hellmut Wilhelm and Richard Wilhelm, 20–32. Princeton: Princeton University Press, 1995.

Wilhelm, Richard. *Die Seele Chinas*. Frankfurt am Main: Insel-Verlag, 1980.

Wilhelm, Richard. *Die Stellung des Konfucius unter den Repräsentanten der Menschheit: Vortrag, gehalten in der Deutschen Kolonialgesellschaft Abteilung Tsingtau*. Tsingtau: Deutsch-Chinesische Verlagsanstalt, 1903.

Wilhelm, Richard. *Kungfutse: Gespräche (Lun Yü)*. Jena: Eugen Diederichs, 1910.

Wilhelm, Richard. *Li Gi: Das Buch der Sitte des älteren und jüngeren Dai*. Jena: Eugen Diederichs, 1930.

Wilhelm, Richard. *The I Ching or Book of Changes*. Translated by Cary Baynes. New York: The American-Stratford Press, 1950.

Wilhelm, Richard and Carl Gustav Jung. *The Secret of the Golden Flower*. Translated by Cary F. Baynes. London: Lund Humphries, 1931.

Wilhelm, Richard. *Tao Te King: Das Buch des Alten vom Sinn und Leben*. Jena: Eugen Diederichs, 1911.

Wilhelm, Richard. *The Soul of China*. Translated by John Holroyd-Reece. New York: Harcourt, Brace and Company, 1928.

CHAPTER 7

Good migrations?

Harry Martinson's travel writing in an age of climate change, refugee crisis and pandemics

Andreas Hedberg
Uppsala University

The paragon of the Swedish author (and later Nobel prize laureate) Harry Martinson's early travel writing, published in the 1930s, is the "geosopher," a figure that has broken free of the confines of his birth nation and its culture in order to experience the world in its entirety. For the geosopher, travelling is a basic need.

Martinson's ideal, based on his own experiences as a ship stoker, was made possible by modern transportation, technology, commerce and cultural transfer on a new scale – phenomena that for some of his contemporaries seemed more frightening than promising, uprooting them from traditional life. Today, in the wake of climate change, the fear of pandemics and large-scale migration, these anxieties have resurfaced.

Even though Martinson seemed optimistic about his geosopher ideal, he soon gave up his life as a migrant and turned to the Swedish countryside, where the global perspective of his travel writing was replaced by an interest in the smallest creatures and movements of nature. When he later returned to the theme of travelling, it was with a completely different tone in the dystopic fantasies of the "space epic" *Aniara.*

In this article, I will explore the shifting attitudes in Martinson's travel writing, and also relate them to our contemporary challenges: migration, climate change and the Covid-19 pandemic. How can literature contribute to cultural transfer and border crossings in an age where mobility has to be limited?

Keywords: knowledge production, mobility/migration/globalisation, identity construction (the self vs the other), space and time

https://doi.org/10.1075/fillm.20.07hed

Our contemporary world is globalised. Capital has gone through several stages of internationalisation. For those who have the means, long-distance travel is easy. Consumer products are transported all over the world. For decades, communication has gone through a process of digitalisation. Yet, despite this development going on for quite some time – which should have given us the chance to mentally catch up – we still see numerous testimonies to a feeling that everything is happening too fast, that life is continuously speeding up and spreading out, as if a powerful centrifugal force is affecting our existence. This feeling – present for example in the growing *prepper* and *off-grid* cultures in the US and Western Europe – may be seen as the result of what Karl Marx called "the annihilation of space by time," a concept further elaborated on by geographer David Harvey, who in his 1989 book, *The Condition of Postmodernity*, speaks of "time-space compression."[1]

This development, which disconnects us from our origins in premodern society, from what Henri Lefebvre, in his book, *The Production of Space* (1991), calls absolute space (a notion very much influenced by Friedrich Nietzsche), has mostly been regarded as a process of emancipation.[2] However, there have also been and still are forces working in the opposite direction. Populist political parties, encouraged by events such as the 2015 so-called "refugee crisis" in Europe, are working to limit the freedom of mobility. The climate movement, a result of the growing awareness of the fatal consequences of carbon emissions and global warming, is demanding strong action against the effects of consumerism and long-distance travel – the "Let's stay grounded" campaign and the Swedish concept of *flygskam* ("flight shame") are clear examples of this trend. Fear of pandemics, such as the 2020 global outbreak of the Covid-19 virus, has equally led to a questioning of the ever-increasing mobility of modern life, including social media shaming travel and large gatherings of people.

Taken together, these movements, whose ideas have also been expressed in dystopic cli-fi narratives, are powerful centripetal forces working against globalisation, resulting in a complicated modern existence and *zeitgeist* which revives questions posed by the social scientist and geographer Doreen Massey in her 1991 essay, "A Global Sense of Place": How, in the context of all these time-space changes – which, to further complicate matters, are socially varied (e.g. as a result of class, gender and ethnicity) – do we think about "places"? and, in an era of globalisation, migration and international capital, how do we think about "locality"?

---

1. David Harvey, *The Condition of Postmodernity: An Inquiry into the Origin of Cultural Change* (Oxford: Blackwell, 1990).

2. Henri Lefebvre, *The Production of Space*, trans. Donald Nicholson-Smith (Cambridge Massachusetts: Blackwell, 1991), 22.

The aim of this chapter is to discuss how processes of globalisation and anti-globalisation have been mirrored in literature during the twentieth century, while also taking into consideration how literature – despite today's questioning of geographical mobility – might contribute to cultural transfer and communication as well as point to ways of coping with time-space compression and isolationism. I will do this through an analysis of the travel writing – both enthusiastic and dystopic – of the Swedish Nobel Prize laureate Harry Martinson (1904–1978), applying a theoretical framework mainly inspired by Massey, but also by the literary scholar Franco Moretti and the ecological thinker Timothy Morton. In order to more fully discuss Martinson's travel writing in relation to modernity and modernisation, I will also make some references to the work of Lefebvre and to the philosopher and literary scholar Marshall Berman. This type of theoretical framework has not yet been used in order to discuss Martinson's work. But as will become clear from my analysis, Massey and Morton especially makes it possible to discover new aspects of his travel writing. The main question posed in this article is this one: how can these new aspects of Martinson's work help us understand and grapple with contemporary challenges with regard to mobility and social development?

As the cultural and historical geographer Peter Merriman has pointed out in his book, *Mobility, Space and Culture* (2012), by providing a general overview of the "mobility turn" in the social sciences, Massey's key aim was to "reconnect accounts of the spatial with the political as well as the temporal," while also incorporating "diverse approaches to space and time in human and physical geography."[3] Merriman goes on to question "the persistent positioning of space and time as primordial and a priori concepts for understanding the unfolding of events *above* other concepts", a positioning based on the writings of an array of processual and post-structuralist thinkers and still present in Massey's work. Being mindful of the risk of reproducing such post-structuralist clichés, I will combine Massey's sociological thinking on migration and cultural mobility with other perspectives. Franco Moretti offers an example of how the concepts of space and time might be applied in literary study (tracing the "*particular* events" and the "*particular* social material assemblages" favoured by Merriman), and Timothy Morton offers a way to engage with the climate crisis, an issue very much relevant to a contemporary discussion of literature, time and space.[4]

After a brief introduction to Harry Martinson's career, I will move on to a chronological discussion of his travel-themed writing and its shifting moods,

3. Peter Merriman, *Mobility, Space and Culture*, International Library of Sociology (Abingdon: Routledge, 2012), 38.

4. Merriman, *Mobility, Space and Culture*, 41.

using the optics created by the fusing of theoretical work by Massey, Moretti and Morton. The article closes with sections relating Martinson's writings to the twenty-first-century renaissance of dystopic narratives, as well as to today's globalist and anti-globalist movements.

Catching the dewdrops, reflecting the cosmos: The work of Nobel Prize Laureate Harry Martinson

The fifth Nobel Prize in literature to be awarded to a Swedish writer, and the third to be awarded to a member of the Swedish Academy itself, was jointly received by the academy members Eyvind Johnson (1900–1976) and Harry Martinson (1904–1978) in 1974. They both belonged to a generation of "proletarian writers," literary autodidacts who rose to fame in the 1930s. Both Johnson and Martinson can also be described – somewhat surprisingly considering their humble beginnings and their base in a vernacular proletarian context – as cosmopolitan writers, having spent most of their youth abroad in order to escape poverty and abuse, Johnson in the literary metropoles of Berlin and Paris, and Martinson as a ship stoker and self-proclaimed "geosof" ("geosopher") or citizen of the world.

These nomadic existences are mirrored in the international transmission of their work, both having been translated into world languages – English and French – very early on in their careers, even before Danish, Norwegian and other Nordic languages, thus reversing the traditional bibliomigration patterns of Swedish writers (as well as the international careers of contemporary and comparable "proletarian writers" such as Ivar Lo Johansson and Vilhelm Moberg), who tend to depend on Danish and German as "transit languages" on the path to international fame. Martinson, whose work will be the focal point of this article, received the Nobel Prize for writings "that catch the dewdrop and reflect the cosmos."[5] This characterisation by the prize committee reflects the wide range of Martinson's career as a writer. Although his depictions of nature and society are well anchored in the soil of his home country, Martinson was also a world trav-

5. Concerning the prize motivation, see "Harry Martinson: Catching the Dewdrop, Reflecting the Cosmos". NobelPrize.org. Nobel Prize Outreach AB 2024. Thu. 30 May 2024. <https://www.nobelprize.org/prizes/literature/1974/martinson/article/> On the concept of bibliomigration, see Yvonne Lindqvist, "Bibliomigration från periferi till semi-periferi. Om den samtida spansk-karibiska litteraturen i svensk översättning," *Tidskrift för litteraturvetenskap*, no. 1 (2018); and Venkat Mani, *Recoding World Literature: Libraries, Print Culture, and Germany's Pact with Books* (New York: Fordham University Press, 2017), 33.

eller and something of a literary astronomer, looking inward, into the microcosmos of life on earth and at the same time outward, in dizzying visions of space.

Today, Harry Martinson is perhaps best known for his poetry – especially for the dystopian "space epic" *Aniara. En revy om människan i tid och rum* (*Aniara: A Review of Man in Time and Space*, 1956) – and for his socially engaged novels, such as the semi-autobiographical *Nässlorna blomma* (*Flowering Nettle*, 1935). His breakthrough in the very first years of the 1930s was, however, mostly due to his expressive depictions of life as a seaman, in both poetry and prose. His cosmopolitan outlook, combined with his fondness for a direct and basic language – often lacking traditional metaphor – made him appear youthful and energetic to the literary establishment. Consequently, he was associated with the Swedish primitivist movement – noted by contemporary critics for its vitalism and free verse – which was established with the anthology, *5 unga* ("Five Young Ones," 1931) and to which Martinson contributed.

Harry Martinson 1932, Aged 28. Photo: Anna Riwkin

Aimless journeys: Rarly 1930s travel writings

The language of Martinson's travel writing has been described as "nytvaget" ("newly washed") and "lysande av spänst och fräschör" ("brilliant in its elasticity and freshness").[6] While referring primarily to the prose collections of *Resor utan mål* ("Aimless Journeys," 1932) and *Kap Farväl!* (*Cape Farewell*, 1933), this characterisation is also true of some of the poems in Martinson's first collections of poetry, *Spökskepp* ("Ghost Ships," 1929) and *Nomad* ("Nomad," 1931). "On the Congo," the introductory poem of *Resor utan mål*, thematically reminiscent of Joseph Conrad's novella *Heart of Darkness* (1899), may serve as an illustrative example:

> Our ship *The Sea Forge* swung out of the trade wind
> and crept up the Congo River.
> The lianas drooped dragging on the deck like log-lines.
> We met the Congo's famous iron barges,
> their hot plate-decks milling with negroes from the tributaries.
>
> Hands raised to mouths
> they called "Devil take you" in a Bantu tongue.
> We slid, wondering and oppressed, through tunnels of greenery
> and the cook in his cabin thought:
> "Here's me peeling spuds in the Congo interior."
>
> By night *The Sea Forge* stared
> with red eyes into the jungles,
> a beast bellowed, a jungle rat plopped in the river,
> a millet mortar coughed hoarsely
> and a dull drumbeat resounded somewhere from a village
> where rubber negroes lived their slave-lives.[7]

The hero of Martinson's 1930s travel writing is the "världsnomad" ("nomad of the world"), a figure that has broken free of the confines of their birth nation and its culture in order to experience the world in its entirety. For the "världsnomad," Martinson explains, travelling is a fundamental need. This ideal of a nomadic existence, based on the writer's own experiences as a seaman, was, of course, made possible by modern transportation, technology, cultural transfer and com-

6. Harry Martinson, *Resor utan mål* (Stockholm: Bokförlaget Aldus/Bonniers, 1967), back cover. Translations of passages in *Resor utan mål* are mine, unless otherwise stated.

7. Martinson, *Chickweed Wintergreen: Selected Poems*, trans. Robert Fulton (Tarset: Bloodaxe, 2010), 47.

merce – phenomena that for some of Martinson's contemporaries seemed more frightening than promising, in uprooting them from traditional life.

In *Resor utan mål*, this ambivalence felt towards movement and travel is partly explained with reference to our common past as a species. For thousands of years, Martinson claims, the "nomadic instinct" of human beings has been repressed. The basis of this repression is to be found in what Timothy Morton has called "agrilogistics," a programme or worldview that has shaped human existence since the Neolithic revolution, in the transition from hunting and gathering to agriculture and settlement around 10,000 BC. According to Morton, the "agrilogistic algorithm" consists of "numerous subroutines," among which are the elimination of "contradiction and anomaly," the establishment of "boundaries between the human and the non-human" and the maximisation of "existence over and above any quality of existence."[8]

All these subroutines are challenged in Martinson's travel writing, which instead emphasise humanity's "innersta vaganta längtan" ("innermost longing for vagrancy"). Consequently, there is an uncanny character (in the Freudian sense of the term) to the ideal of the "världsnomad," since it reminds us of a repressed (pre-Neolithic) past that has been reawakened by the Technological Revolution of the late nineteenth and early twentieth centuries (also known as the Second Industrial Revolution). This uncanniness is addressed head on in *Resor utan mål.* From the point of view of the "kälkborgare" ("the philistine"), Martinson claims, travel writing is dangerous because it produces longing: "[a]llt som doftar nomad är *den ordentligt förträngdes fasa*, det är för honom något oanständigt och asocialt" ("everything that smells of nomad is to the horror of the properly repressed, it is for him something obscene and antisocial"). Martinson, a believer in "motsatsens vaganta religion" ("the opposite religion of vagrancy"), considered it his mission to tenaciously break through to what might be termed the "post-agrilogistic" phase of history, "nomadframtiderna" ("the nomadic futures").[9] Nevertheless, there remains a sense of ambiguity in Martinson's travelogues, as if there is a constant need to overcome a sense of fear and insecurity.

This sense is typical of early twentieth-century Swedish literature, where a number of authors used their work to elaborate on a feeling of loss that seems to have been intrinsically linked to the modernisation of society during the Technological Revolution. Among them were Sweden's first Nobel Prize laureate, Selma Lagerlöf (1858–1940). Mourning the loss of her childhood home of Mårbacka in the province of Värmland, Lagerlöf turned her personal grief into a melancholic

8. Timothy Morton, *Dark Ecology: For a Logic of Future Coexistence* (New York: Columbia University Press, 2016), 46.

9. Martinson, *Resor utan mål* (Stockholm: Bonnier, 1932), 7–9.

vision of contemporary humanity, being forcefully uprooted from traditional rural existence and thrown into what Marshall Berman, in his seminal book, *All That Is Solid Melts into Air* (1982), has described as "the maelstrom of modern life," where the industrialisation of production "creates new human environments and destroys old ones."[10] New technology meant new possibilities to transport people and goods but also to transfer ideas, political opinions and aesthetic movements, as Orlando Figes demonstrates in his study, *The Europeans: Three Lives and the Making of a Cosmopolitan Culture* (2019). This meant a cosmopolitanisation of European society, which could be seen as a threat to traditional ways of life, rooted in the vernacular. Consequently, the technological revolution was accompanied by an increase in nationalist feelings, anti-cosmopolitanism being one of the reasons for the outbreak of the First World War.

The cognitive and existential consequences of these processes were analysed by Franco Moretti in his book, *Graphs, Maps, Trees* (2005), as a shift from a circular to a linear worldview that was fuelled by parliamentary enclosure laws, where a perceptual system in which the rural village is "still largely self-sufficient and can therefore feel at the centre of 'its own' space" is replaced by "an abstract grid," within which the village becomes "just one of the many 'beads' that the various roads will run through."[11] This shift from a circular to a linear perception of the world is an important theme in Lagerlöf's novels, *Jerusalem* (*Jerusalem*, 1901–1902) and *Kejsarn av Portugallien* (*The Emperor of Portugallia: A Story from Värmland*, 1914), in which the main characters are forced to leave rural villages to confront the urban landscapes of modernity in Sweden and the world. The same might be said about novels by some of Lagerlöf's contemporaries (and compatriots), widely read at the time, but today more or less forgotten, such as Karl-Erik Forsslund (1872–1941) and Sven Lidman (1882–1960).[12]

Martinson's travel writing addresses the same issues as Lagerlöf's novels, but does so in the light of experiences undergone by a later generation. In the 1930s, there no longer seems to be any place for nostalgic renderings of rural life. Rather, Martinson and his contemporaries stressed the need to push on towards the future, with their vision of modernity echoing Berman in his analysis of Johann

10. Marshall Berman, *All That Is Solid Melts into Air: The Experience of Modernity* (New York & London: Penguin Books, 1988), 16.

11. Franco Moretti, *Graphs, Maps, Trees: Abstract Models for Literary History* (New York & London: Verso), 2007, 38.

12. Cf. Andreas Hedberg, "Lyckan heter hem. Den hotade gården som motiv i svenska romaner 1899–1915," in *Spänning och nyfikenhet. Festskrift till Johan Svedjedal*, ed. Gunnel Furuland, Andreas Hedberg, Jerry Määttä, Petra Söderlund & Åsa Warnqvist (Möklinta: Gidlunds, 2016), 56–67.

Wolfgang von Goethe's *Faust* – "the crucial thing is to keep moving."[13] Consequently, Lagerlöf's and Forsslund's depictions of the rural farm as a metaphor for the nation were no longer valid. Since the modernisation of European society, spurred on by the Enlightenment and the subsequent waves of industrialisation, had broken "the primacy of the spatial," as described by Lefebvre in *The Production of Space*, common ground had to be found in new places, or rather in new forms of existence, where the absolute space of Nietzschean philosophy no longer dictated humanity's possibilities.[14]

For Martinson, modern existence is about escaping from the "staketinhägnade" ("fenced-in") emptiness of "jäktet" ("the hustle"), the limitedness of the old-fashioned nations and daily work, and transitioning to "ett levande relationsliv" ("a vivid life of relations"). Life in the twentieth century, he claims, comes with never previously seen possibilities for "psyko-biologisk" ("psycho-biological") recreation, with new conditions and new horizons, making the modern human being a "frigjord förvaltare och intensifiator av sitt eget psyke" ("liberated captain and intensifier of his own psyche").[15] He thought that the ambiguity felt by Lagerlöf's generation, but also by his contemporaries, and to some extent also by himself (judging by the tone in some of his travelogues), in the face of modernity and its accelerated communications, must and will be overcome: "[f]ramtidens allkommunicerande världsfolk [...] skall en gång [...] skratta gott åt Undergångens katalog, och åt kulturerna som inte vågade upplösa sig (specieförändras) och 'gå under'" ("[o]ne day, the all-communicating world-people [...] of the future will laugh at the Catalogue of Doom, and at the cultures that did not dare to dissolve (change species) and 'perish'").[16]

Butterfly, moth and crane fly: Late 1930s nature writing

The extraordinary self-confidence of Martinson's early 1930s travelogues was, at least to some extent, a rhetorical strategy. The insecurity and unsettling impact of time-space compression was predestined to strike back, mirrored in the fact that Martinson, having contracted tuberculosis, had to come ashore for good in 1927. Consequently, when writing about his "geosopher" ideal in the early 1930s, he was already unable to *live* it.

13. Berman, *All That Is Solid Melts into Air*, 50.

14. Lefebvre, *The Production of Space*, 22.

15. Martinson, *Resor utan mål* (1932), 84–5.

16. Martinson, *Resor utan mål* (1932), 86.

During the second half of the decade, Martinson took a step back from the global perspective, retreating to family life in the countryside of southern Sweden. In his writings, he turned to lyrical miniatures depicting the details of nature, a trend already visible in the collection *Natur* ("Nature"), published in 1934. The short poem "Fjärilen" ("The Butterfly") is a good example of this motif in Martinson's writing:

> Born to be a butterfly
> my cool flame flickers
> in the heavy velvet of the grass.
> The children chase me. The sun goes down
> beyond the mallows and the tussock,
> rescuing me till nightfall.
> The moon rises: it's far away, I'm not afraid,
> I listen to its beams.
> My eyes film over protectingly.
> My wings are stuck together by dew.
> I sit on the nettle.[17]

Natur was published to mixed reviews, perhaps proving that this period of urbanisation and technological revolution in Sweden was not the best time to praise and protect nature. In the following year, Martinson made his debut as a novelist, with the semi-autobiographic *Nässlorna blomma* (*Flowering Nettle*). In these thinly veiled childhood recollections that trace the life of the main character, Martin, until the age of thirteen, Martinson reveals himself to be a writer who was quite preoccupied with the origins that, until then, he had been so eager to escape. It seems that the trauma of growing up as an orphan finally caught up with him and had to be dealt with. *Nässlorna blomma* is the story of an orphan who finally breaks free to become an independent citizen of the world, but it is a story written by a man who, in the process of writing, had to force himself to re-experience his hardships.

In the very last years of the 1930s, Martinson once again changed his style of writing, publishing three collections of prose that can most aptly be described as nature writing in the vein of Henry David Thoreau. In these meditations, inspired by life in the countryside just south of Stockholm, Martinson seems to claim that humanity needs nature much more than nature needs humanity, continuing themes expressed in the *Natur* poem "Kraft" ("Power"), where he speaks of "boastful cities," in contrast to the quiet life in the countryside that feeds them.[18]

17. Martinson, *Chickweed Wintergreen*, 52.
18. Martinson, *Chickweed Wintergreen*, 51.

In other words, humanity has been shaped by evolution to live in close correspondence with nature, and not in the urban environments of modern cities. This could be seen as a step back from the travelogues of *Resor utan mål*, and a return to the critique of modernity and civilisation common among Martinson's predecessors, writers such as Selma Lagerlöf, Karl-Erik Forsslund and Sven Lidman.

For Martinson, modernity was now also closely tied to the totalitarian movements of the 1930s. The war-mongering of the German Nazis was a threat not only to the lives of innocent people, but also threatening the destruction of the very living conditions of human beings. In the first of his books of nature writing, *Svärmare och harkrank* ("Moth and Crane Fly"), Martinson denounces "vår tids heroistiska rörelser" ("the heroism movements of our time"), explicitly mentioning the Italian fascists as an example. Interestingly, his critique takes the form of a condemnation of the same type of decoupling that he proclaimed as an ideal in *Resor utan mål* and *Kap farväl!*. The heroism movements, according to Martinson at the time, are characterised by self-glorifying myths and artificial attempts to "make history" by pulling modern humanity "out of the 'ordinary'."[19] This "making of history" is in fact one of the fundamental myths of modernity analysed by Berman through Goethe's *Faust*, and described by Lefebvre as a move from absolute to abstract space. What is interesting here, however, is that Martinson, after witnessing the growth of totalitarian movements during the 1930s, appears to shift his position, becoming less of a champion and more of a critic of the ongoing transformation. Italian Fascists and German Nazis, he seems to say, are abusing and corrupting the achievements of modern humanity.

A dark and icy wind: The space Epic *Aniara*

Martinson's sorrow over the repeated tragedies of the twentieth century is, however, most poignantly expressed in the space epic, *Aniara* (1956). The background to this work of 103 cantos, written in a variety of meters and with a multitude of neologisms, is the specific *zeitgeist* of the immediate post-war period, characterised by a great anxiety after the first use of atomic bombs, which seemingly opened the gate to humanity's self-inflicted extinction. The spaceship Aniara, escaping from the radioactively infected earth, has gone off course and drifts helplessly towards the constellation of the Lyre. The wonderful Mima, the ship's conscious super-computer, with superhuman artificial intelligence, dies of grief ("[d]arkened in her cellworks by the cruelty/man exhibits in his hour of sin") over

19. Harry Martinson, *Svärmare och harkrank* (Stockholm: Bonnier, 1937), 32–3. Translations of passages in *Svärmare och harkrank* are mine.

Illustration for Harry Martinson's *Resor utan mål,* Bertil Bull Hedlund

humanity's folly and collective suicide.[20] The indifference, grief and final annihilation of the passengers only repeat what has happened on earth.[21]

To some extent, *Aniara* is the exact opposite of *Resor utan mål.* Martinson has returned to the theme of travelling, but the journey has become interstellar and tarnished with the sins of humanity. The decoupling of humanity from its origins

20. Harry Martinson, *Aniara: A Review of Man in Time and Space*, trans. Stephen Klass & Leif Sjöberg (Södra Sandby: Vekerum, 1991), 47.

21. For this description of the main events in *Aniara*, I am indebted to the literary scholar Göran Hägg and his account of Martinson's space epic in the book, *Den svenska litteraturhistorien* (Stockholm: Wahlström & Widstrand, 1996).

has gone through a process of change, evolving from a great promise to the fundamental tragedy of modern existence. In *Aniara*, travelling is sorrowful (the name Aniara is based on the Greek word ἀνιαρόζ, meaning grievous or painful), something to which we are condemned. Much like the spaceship's passengers, 20th century humanity is drifting aimlessly, deracinated from its geographical, social and cultural environment. Realising that their drift is "even deeper" than they first believed, that they have "strayed in spiritual seas," Aniara's passengers burst into nostalgic dirges about life on earth.[22] A blind woman, having witnessed the complete nuclear destruction of her homeland ("hearing everything I loved/vanish in a dark and icy wind"), seeks the singer's job in Aniara's Chamber Three, where she performs songs paying homage to nature on Earth ("Ah, the Dale, ah Me" and "Little Bird out in the Rosewoods Yonder").[23] These songs can easily be interpreted as channelling the passengers' – as well as Martinson's and his readers' – longing for enrootedness, for an absolute space (in the Lefebvrian sense), to which they are inexorably unable to return. Paradoxically, this also means a longing for the very space that Martinson in his travelogues was so eager to leave behind.

Twenty-first-century Renaissance of dystopic narratives

One fascinating thing about Martinson's writing is that it is so versatile, encapsulating several important attitudes to the world and to the processes that had transformed the conditions of human existence since the beginning of the twentieth century. Consequently, it can be used as an intellectual toolbox while trying to come to grips with existence, in what Berman termed "the maelstrom of modern life." When writing this article, in late 2020, our existence seems to have been overturned by the centripetal, anti-globalist forces described in the introduction to this article. The Covid-19 pandemic has made international and intercontinental journeys difficult or even impossible. Even in pre-pandemic society, the climate crisis had put a damper on long-distance travel, with its contribution to carbon emissions, while the "refugee crisis" of 2015 had given rise to fear-mongering political movements calling for closed borders, fences and walls.

This seems very much like a return to the pre-globalist nationalisms and nostalgia at the centre of literary works by Martinson's predecessors, suggesting that the era of the "geosopher" or "the nomad of the world" (so confidently inaugurated in the travelogues of *Resor utan mål* and *Kap farväl!*) was a historical

22. Martinson, *Aniara*, 25.

23. Martinson, *Aniara*, 82–3.

anomaly, limited to what the British historian Eric Hobsbawm has termed "the short 20th century" (1918–1989). This "short century" seems in many ways to be a period of anomalies. In his book, *Le Capital au XXIe siècle* (*Capital in the Twenty-First Century*, 2013), the French economist Thomas Piketty shows that the general trend towards greater inequality, a fundamental feature of capitalist economy, was reversed between 1930 and 1970 due to unique historical circumstances. As a result of leaving the anomalies of the "short century" behind, it seems that contemporary readers find it easier to feel a kinship with the pessimism of *Aniara* than with the triumphant spirit characterising the travelogues of the early 1930s. Today's great success of dystopic narratives (as shown by the intermedial and international trajectories of Cormac McCarthy's *The Road*, Margaret Atwood's *The Handmaid's Tale* and Aldous Huxley's *Brave New World*) is convincing proof of this. It should thus come as no surprise that *Aniara*, only in the last decade, has been made into a musical (in 2010), a graphic novel (in 2015) and a feature film (in 2019).

Dystopic thinking readily lends itself to conservative ideals, reverting to protectionism and xenophobia in the face of a technical and societal development (thematised and denounced in dystopic narratives) that comes across as frightening and destructive. Today, we are not only dealing with the climate crisis, pandemics and war-induced migration, but also with what the German economist Klaus Schwab has called the Fourth Industrial Revolution – the ongoing automation of traditional manufacturing and industrial practices based on large-scale machine-to-machine communication and the "internet of things."[24] Space-time compression, or the annihilation of space by time, is once again speeding up, evoking experiences comparable to those felt by Selma Lagerlöf and her contemporaries, who were trying to come to terms with the effects of the Second Industrial Revolution at the turn of the twentieth century.

It seems natural, then, that the critique of civilisation present in the work of Lagerlöf reappeared in contemporary Swedish literature during the 2010s. This is evident in the writings of authors such as Jesper Weithz (b. 1974) and Helena Granström (b. 1983), who, in poetry, prose and essays are questioning modern technology and its effect on nature and the human psyche.[25] Once again, time, as a driving force of existence, is disputed. Instead, there is a renaissance of spatiality and a call to re-embed humanity in a Lefebvrian absolute space. Thus, the partic-

24. Klaus Schwab, *The Fourth Industrial Revolution* (New York: Crown Business, 2017).

25. Andreas Hedberg, "Mänskligheten svämmar över alla bräddar och krymper på samma gång. Modernitet och antropocentrism i Karl-Erik Forsslunds *Djur* (1900) och Helena Granströms *Det som en gång var* (2016)," *Tidskrift för litteraturvetenskap*, no. 1 (2019), 60–9.

ular physical place, the *locale*, is assigned crucial importance – as a home, but also as something which should be allowed to shape human existence.

According to Lefebvre, this Nietzschean perception of time and space differs significantly from its Marxist counterpart.[26] In the twentieth century, Nietzsche's vision was further developed by Martin Heidegger, who, in his essay "Bauen Wohnen Denken" ("Building Dwelling Thinking," 1951), speaks of "bring[ing] dwelling into the fullness of its essence"[27] or to "preserve the fourfold," which means "to save the earth, to receive the sky, to await the divinities, to initiate mortals."[28] This, of course, is a tempting strategy, since a strong sense of place can offer peace and security; a refuge from the maelstrom of time-space compression. However, as was made evident through Heidegger's inclination for the national socialists, this perception of spatiality and place easily taps into the same thinking that produced the *heimat* movement and the ideals of *blut und boden*. This is one of the reasons why today's idea of place as a guiding principle for human existence must be rethought as something socially and ecologically sound. I would like to end this article with a discussion of this other type of place – inspired by Massey's idea of a "global sense of place" and Morton's of a "logic of future coexistence". This will also be a discussion of Martinson's travel writing, which is what makes it possible to bring these theoretical concepts together.

A vivid life or relations: Martinson's global sense of place

In *Dark Ecology*, Morton picks up the thread of place and spatiality in the wake of the climate crisis. Since the 1970s, he claims, "[m]any have pronounced the death of the place," the death of the localised and particular. In literary studies, this has been done hand in hand with "the language of textuality versus speech." However, according to Morton, the opposite has occurred; what has collapsed is instead "(the fantasy of empty, smooth) space," which has revealed itself as an "anthropocentric concept," as "the convenient fiction of white Western imperialist humans."

This, however, does not mean that we should retreat to the localised and particular. Nostalgia for the old ways is as fictional as modernity's celebrations of deracination. For Morton, the ecological era is "the revenge of place, but it's not your grandfather's place" (and not, incidentally, "the Aryan homeland," since

26. Lefebvre, *The Production of Space*, p. 22. Cf. Hedberg, "A Living Home," 231.

27. Martin Heidegger, "Building Dwelling Thinking," in *Basic Writings*, ed. David Farrell Krell (San Francisco: HarperCollins, 1993), 343–64, here: 363.

28. Heidegger, "Building Dwelling Thinking," 360.

"[t]he very idea of points of origin is an agrilogistic hallucination").[29] Place has nothing to do with "good old reliable constancy," as there are too many "intersecting places," too many scales and too many "nonhumans" to be taken into consideration. When massive "objects" such as global warming or humanity on a global scale are made "thinkable," they grow near, changing the idea of place, even though they are so massive that they are impossible to grasp empirically. As "hyperobjects," they remind us that the local, or place, is in fact the uncanny – intimate and monstrous at the same time.[30] In order to co-exist with this kind of place in the future, we have to rid ourselves of the agrilogistic algorithm as well as our anthropocentric privileges, in order to be part of what Morton calls the *mesh*, a "sprawling network of interconnection without centre or edge."[31]

Morton's thinking on ecology is to some extent analogous with Massey's on migration, cultural mobility and social geography. While Morton is responding to the climate crisis, Massey was responding to the digital revolution (termed the Third Industrial Revolution by some) and the constant movement and intermixing of "the global village," a concept that occupied many in the early 1990s. Having noticed that an "(idealised) notion of an era when places were (supposedly) inhabited by coherent and homogenous communities" is set against "the current fragmentation and disruption," sometimes giving rise to "defensive and reactionary responses," Massey asks herself how a sense of place can be reinvented as something progressive – not "self-inclosing and defensive, but outward-looking."

Much like Morton imagines a place without "good old reliable constancy," a place as an intersection of a multitude of phenomena on different scales, Massey puts forward her idea of a "global sense of place."[32] Instead of thinking of places as areas with boundaries around them, they might be imagined as "articulated moments in networks of social relations and understandings, but where a large proportion of those relations, experiences and understandings are constructed on a far larger scale than what we happen to define for that moment as the place itself." This, in turn, allows for "a sense of place which is extroverted, which includes a consciousness of its links with the wider world," for example through the cultural transfer of literature, and "which integrates in a positive way the global and the local."[33]

---

29. Morton, *Dark Ecology*, 66.

30. Morton, *Dark Ecology*, 9–11. Cf. Morton, *Hyperobjects: Philosophy and Ecology after the End of the World*, Posthumanities, 27 (Minneapolis: University of Minnesota Press, 2013).

31. Morton, *Dark Ecology*, 81.

32. Doreen Massey, "A Global Sense of Place," *Marxism Today* 38 (June 1991), 24–9, here: 24.

33. Massey, "A Global Sense of Place," 28.

For Morton, entering the post-agrilogistic is a painful process; it means meeting the uncanny (a tension that is vividly thematized in Martinson's *Resor utan mål*). However, to meet the uncanny is also to come home (much like the word "unheimlich," according to Freud, exhibits a meaning which is identical with its opposite, "heimlich"); it is returning to what Morton calls the *arche-lithic* and its ecological awareness (*ecognosis*), "a primordial relatedness of humans and non-humans that has never evaporated." The agrilogistic is all about maintaining control through inventing and objectifying the "non-human." This control has to be relinquished to make possible a future coexistence of the kind that Morton imagines.

The Technological Revolution of course represents an important challenge to agrilogistic thinking, since it uprooted so many from traditional agrarian life. As has been shown above, this deracination is also, at least to some extent, what Martinson embraced in his 1930s travel writing. The "geosopher" must abandon control over his "grandfather's place" – over place as something localised and particular that serves humanity by producing its livelihood – in order to experience space in a wholly different way, as a dizzying multitude of places that intersect in his psycho-biologically changing psyche. Travelling, however, is never abstract, never quasi-fictional, as are today's trips on intercontinental flights. Travelling is *felt* as something physically affecting the body, which is evident in this passage from *Resor utan mål*:

> To travel, I have found, is – on a hot Pomeranian summer day – to shovel coal upstream on the river Oder, to stick your head out of the stokehold every three minutes to overlook the Pomeranian plain in a geographical moment bought with blood and sweat, and then to dive down again.[34]

When travelling is felt like this, it does not mean simply abandoning the "good old reliable constancy," nor is it an unproblematic celebration of uprootedness. Rather, it concerns accepting our connection to the non-human world, while at the same time reshaping this connection, freeing it from the simplistic power dynamic and objectification of Neolithic agrarian life. It is, at least to some extent, taking part in the *mesh*.

However, Martinson's challenge to agrilogistic existence does not end there. He also questions, often through metaphor, the anthropocentric dimension of the Neolithic worldview as such. In *Resor utan mål* and *Kap farväl!*, he makes frequent use of anthropomorphisms ("the iron belly of the boat," "the laughter of the Atlantic"),[35] blurring the boundary between human and non-human. Travelling,

34. Martinson, *Resor utan mål* (1932), 10.

35. Martinson, *Resor utan mål* (1932), 33 & 48.

with its many changes in experiences and perspectives, also seems to foster or facilitate the non-anthropocentric coexistence favoured by Morton, as in this description of one of the waterfalls in the province of Santa Catarina, Brazil:

> I sat down on a relatively dry spot to look and listen. I was quiet and big-eyed and watched magic. Hour after hour passed, and hour after hour I sat. This I have done at every waterfall I have visited. I simply sit, and the great water-cosmos thunderously sings. I release my spirit from dress and sex and become receptive only – become very nearly a child beside the great roaring waters.[36]

Experiencing the world as a nomad, a traveller characterised by his fundamental disinterest in relation to the world, allows for a stepping out of old destructive roles (species, gender, etc.), opening up to the *arche-lithic* and its *ecognosis*, without – as Morton puts it – "dichotomies of good and evil, need and want, Nature and Culture, human and nonhuman, life and nonlife, self and nonself, present and absent, something and nothing."[37] To "become receptive," in Martinson's sense, means opening the gates to the great non-anthropocentric communion with the world.

If it is possible for us to "become receptive" in relation to our natural environment, it should also be possible to achieve the same state of mind in relation to our social and cultural surroundings. This means breaking free from what Martinson, in *Resor utan mål*, called the "fenced-in" emptiness of the "hustle," or what Massey calls a "self-enclosing" or "defensive" sense of place. Consequently, Martinson's vision of a "vivid life of relations" that would liberate and intensify our psyche is very much comparable with Massey's wish for cultural mobility and a global sense of place.

Conclusion

For the young Harry Martinson, human existence is inextricably linked to travelling, to life as a disinterested nomad. This becomes clear when reading his 1930s writing, both his poetry and his prose travelogues *Resor utan mål* and *Kap farväl!*. And life in our contemporary, digitalised society, with its many examples of cultural transfer and translation, offers ample possibilities to achieve a similar outlook, even though we, as an effect of global warming and pandemics, have been forced to abandon the Faustian principle of movement – at least if this prin-

36. Harry Martinson [author's name misspelt: Harry Martinsson], *Cape Farewell*, trans. Naomi Walford, illust. John Farleigh (London: The Cresset Press, 1934), 113.

37. Morton, *Dark Ecology*, 83.

ciple is interpreted literally. However, as has been shown in this article, the possibility of unlimited movement has also been seen as a threat. Consequently, the nomadic experience of the traveller in Martinson's accounts is very complex. It changes from the perception of nomadic existence as an ideal to the conception of the nomad as a traveller disinterested in the world, longing not for mobility but for attachment, for an absolute space rather than uprootedness. This problematic and conflicted view of mobility was evident already in the early days of modernity, and it still is today – in the minds of individuals as well as in the political landscape.

In her essay "A Global Sense of Place", geographer and social scientist Doreen Massey asks us to "face up to," rather than simply deny, "people's need for attachment of some sort, whether through place or anything else." However, this place does not need to be the product of reactionary nationalism, as in the political programmes of today's populist movements (aiming for "your grandfather's place"), such as the Sweden Democrats in Martinson's home country. Instead, much like Massey proposed in the early 1990s, we need to create a sense of place that will "fit in with the current global-local times and the feelings and relations they give rise to" (e.g. the global village), and that will also be "useful in what are [...] political struggles often inevitably based on place."[38]

One key realisation here is that a place does not have a single, essential identity; a place is always – as it might be put using ecologist thinker Timothy Morton's vocabulary – a vertex where phenomena intersect on many different levels. A place is a process. A place, as Massey puts it, is "a constellation of social relations, meeting and weaving together at a particular locus."[39] Massey's post-agrilogistic vision and Martinson's 1930s travel writing in *Resor utan mål* and *Kap farväl!* – where places are often seen and described in relation to the multifaceted geographical and psycho-biological network of the traveller – offer a modernist conception that we need not abandon. Martinson's work engages with his audience in different ways. He also invites twenty-first-century readers to reflect on contemporary challenges, as they are described in many recent travel narratives. Pandemics, climate change and migration will continue to affect our communities (a situation vividly mirrored in contemporary dystopic narratives in the vein of *Aniara*), and in order to cope with such challenges, we will have to foster a socially and ecologically sound view of life. This may be the "global sense of place," and may also be the "vivid life of relations."

38. Massey, "A Global Sense of Place," 26.

39. Massey, "A Global Sense of Place," 28.

References

Berman, Marshall, *All That Is Solid Melts into Air: The Experience of Modernity*. New York & London: Penguin Books, 1988.

Harvey, David, *The Condition of Postmodernity: An Inquiry into the Origin of Cultural Change*. Oxford: Blackwell, 1990.

Hedberg, Andreas, "Lyckan heter hem. Den hotade gården som motiv i svenska romaner 1899–1915." In *Spänning och nyfikenhet. Festskrift till Johan Svedjedal*, edited by Gunnel Furuland, Andreas Hedberg, Jerry Määttä, Petra Söderlund & Åsa Warnqvist, 56–67. Möklinta: Gidlunds, 2016.

doi Hedberg, Andreas, "Mänskligheten svämmar över alla bräddar och krymper på samma gång. Modernitet och antropocentrism i Karl-Erik Forsslunds *Djur* (1900) och Helena Granströms *Det som en gång var* (2016)." *Tidskrift för litteraturvetenskap*, no. 1 (2019): 60–9.

Heidegger, Martin, "Building Dwelling Thinking." In *Basic Writings*, edited by David Farrell Krell, 343–364. San Francisco: HarperCollins, 1993.

Hägg, Göran, *Den svenska litteraturhistorien*. Stockholm: Wahlström & Widstrand, 1996.

Lefebvre, Henri, *The Production of Space*. Translated by Donald Nicholson-Smith. Cambridge Massachusetts: Blackwell, 1991.

doi Lindqvist, Yvonne, "Bibliomigration från periferi till semi-periferi. Om den samtida spansk-karibiska litteraturen i svensk översättning." *Tidskrift för litteraturvetenskap*, no. 1 (2018): 90–104.

Mani, Venkat, *Recoding World Literature: Libraries, Print Culture, and Germany's Pact with Books*. New York: Fordham University Press, 2017.

Martinson, Harry, *Resor utan mål*. Stockholm: Bonniers, 1932.

Martinson, Harry, *Cape Farewell*. Translated by Naomi Walford, illustrated by John Farleigh. London: The Cresset Press, 1934. [author's name misspelled: Harry Martinsson]

Martinson, Harry, *Svärmare och harkrank*. Stockholm: Bonnier, 1937.

Martinson, Harry, *Resor utan mål*. Stockholm: Bokförlaget Aldus/Bonniers, 1967.

Martinson, Harry, *Aniara: A Review of Man in Time and Space*. Translated by Stephen Klass & Leif Sjöberg. Södra Sandby: Vekerum, 1991.

Martinson, Harry, *Chickweed Wintergreen: Selected Poems*. Translated by Robert Fulton. Tarset: Bloodaxe, 2010.

Massey, Doreen, "A Global Sense of Place." *Marxism Today* 38 (June 1991): 24–9.

doi Merriman, Peter, *Mobility, Space and Culture. International Library of Sociology*. Abingdon: Routledge, 2012.

Moretti, Franco, *Graphs, Maps, Trees: Abstract Models for Literary History*. New York & London: Verso, 2007.

Morton, Timothy, *Hyperobjects: Philosophy and Ecology after the End of the World*. Posthumanities, 27. Minneapolis: University of Minnesota Press, 2013.

doi Morton, Timothy, *Dark Ecology: For a Logic of Future Coexistence*. New York: Columbia University Press, 2016.

Schwab, Klaus, *The Fourth Industrial Revolution*. New York: Crown Business, 2017.

CHAPTER 8

Exile, travel narrative and cultural transfer in Négar Djavadi's *Désorientale* (2016)

Jeanette den Toonder
University of Groningen

This chapter aims to analyse how the intersection of travel and exile in Négar Djavadi's novel *Désorientale* (2016) motivates cultural exchanges through movements of conflict and contact. After having established a framework of exile and travel in relation to narrative, cultural encounters and transfer, the analysis firstly focuses on the metaphorical and physical levels of travel and exile as developed in the novel. Exile and travel are intertwined and a necessary part of the female protagonist's life and offer the possibility to create a travel narrative where she connects her Iranian past and French present. The second part examines the confrontation with her sexual identity and the understanding of her hybrid body as a place of encounter between different cultures. By acknowledging the disorientation of her exiled body, the protagonist/narrator is capable of establishing relationships with others and to share her experience. The final section further discusses the importance of meaningful encounters by elaborating on the connection between the narrator and the implied reader that allow for critical reflection and transfer of insights, emotions and ideas. The analysis demonstrates the role of the exiled narrator as cultural transmitter. This chapter further contributes to the study of travel writing and cultural transfer by offering new perspectives on how the genre develops in the twenty-first century as an effect of migration.

Keywords: Franco-Iranian literature, travel narrative, exile, cultural encounters, sexual identity, disorientation, cultural transfer

Kimiâ Sadr, the narrator of Franco-Iranian author Négar Djavadi's first novel *Désorientale* (2016),[1] tells the saga of her Iranian family and the history of Iran while waiting in a fertility clinic waiting room in Paris for an in vitro fertilisation

1. For the purpose of this chapter I have used the English translation of the novel by Tina Cover. All references are to this translation, *Disoriental* (New York: Europa Editions, 2019).

https://doi.org/10.1075/fillm.20.08den

treatment. She recounts her story non-chronologically and insists on many moments and places in the distant and recent past. Through her narration, she travels back to her birth place, Tehran, and reflects on her childhood memories before and after the Islamic revolution of 1979 and on her family's departure from Iran. The book consists of two parts entitled Side A and Side B, referring to the old 45 rpm vinyl records and the role music played in the protagonist's life as an adolescent. The first part recounts the saga of her family, from the story of her great-grandfather to the budding romance between her parents, their marriage and the birth of their three daughters. Taking on the role of historiographer, the narrator reconstructs this saga in relation to the history of Iran, adding footnotes that refer to important historical events. This story is furthermore intertwined with personal memories of her childhood, and her experience as a displaced girl growing up in Paris. The second part opens with an excerpt of the book her mother wrote when living in France, recounting their flight from Iran from her perspective. This part focuses on how Kimiâ and her family try to adapt to their life in exile, a difficult process that ultimately results in the family falling apart and the assassination of their father by Iranian mercenaries. This part also insists on the protagonist's development from adolescent to young woman, when she decides to leave France and spend several years travelling around the world. During these travels she encounters many different people, one of them is the Flemish girl Anna, with whom she starts a long-term relationship.

In this contribution I will focus on Kimiâ's experience of exile by closely examining her life in France and referring to childhood memories that inform her feelings of alienation, while also discussing the effects of her encounters with others. This second part of the book, which starts with the account of the flight from Tehran and the arrival at Orly airport, is most productive to the study of cultural transfer. It allows for an examination of conflicts and encounters between the protagonist's culture of origin and her newly adopted culture as well as an analysis of her capacity to connect these cultures. The first part of the book, the family saga, precedes the narrator's travel narrative and will therefore not be analysed in this chapter.

Parallel to the story of migration, the question of gender identity is central to the development of the young protagonist, and the experience of exile will be examined in conjunction with gender and sexual development. By intersecting gender, travel account and migration, this chapter particularly contributes to the study of travel writing by offering new perspectives on how the genre develops in the twenty-first century as an effect of migration.

The narrator's story, or rather the variety of stories she tells, take the form of an extended flashback, occurring during the time spent in the hospital waiting room. By constantly moving back and forth between the present and past and between different places and subjects, the narration gives insight into the alienating effects of exile. This feeling of disunity is reinforced through mechanisms of exclusion based on normative gender roles. The title of the novel, combining the two words Oriental ('from the East') and désorienter ('losing one's direction or orientation') expresses the experience of the narrator who crosses geographical borders, eras, and (gender) identities, but also that of the reader whom she invites to join her in this disorienting experience.

Following Stephen Greenblatt's perspective on cultural mobility, in particular the mechanisms of zones of intercultural contact[2] and the sensation of rootedness,[3] this chapter considers metaphorical as well as physical movement of the exiled and disoriented self. These movements are expressed through travel narrative, manifesting itself as a mental and physical plot device which, on the one hand, enables the protagonist to come to terms with her own (up)rootedness, and, on the other, creates zones of contact and exchange. The intersection of travel and exile in Djavadi's novel motivates cultural exchanges through movements of conflict and contact. The travel narrative is characterised by critical moments of leave-taking, remembering home and returning home in narrative, which will be examined within the framework of cultural transfer in which the exiled self is shaped through the encounter of two very different cultures. By taking the perspective of the exiled self as a starting point, this analysis intends to build upon and develop the study of cultural transfer and travel writing through immaterial objects as identified by Walter Moser. Moser argues that cultural transfer takes place when an identifiable agent (or subject) transfers cultural material (or object) from one system to another under concrete historical conditions.[4] One of the categories Moser distinguishes when identifying the objects of transfer is the immaterial object (the other categories are the material and unique object, the serial object, the semiotic or symbolic object and the matrix object). In order to transfer immaterial objects, which Moser identifies as ideas, concepts, values,[5] he argues that material supports are needed. Within the framework of this analysis, the category of immaterial objects enables the study of Iranian and French

2. Stephen Greenblatt, "A Mobility Studies Manifesto," in *Cultural Mobility: A Manifesto*, ed. Stephen Greenblatt, with Ines G. Županov, Reinhard Meyer-Kalkus, Heike Paul, Pál Nyíri, and Friederike Pannewick (Cambridge: Cambridge University Press, 2010), 251.

3. Greenblatt, "A Mobility Studies Manifesto," 252.

4. Walter Moser, "Pour une grammaire du concept de « transfert » appliqué au culturel," in *Transfert: Exploration d'un champ conceptuel*, eds. Pascal Gin, Nicolas Goyer, & Walter Moser (Ottawa : Les Presses de l'Université d'Ottawa, 2014): "il y a transfert culturel quand un agent (sujet) identifiable transfère un matériau culturel (objet) d'un système à un autre dans des conditions historiques concrètes," 53.

5. Moser, "Pour une grammaire du concept de « transfert » appliqué au culturel," 68–69.

values, practices, and characteristics which are supported by the travel narrative and transferred by the narrator of that narrative. The historical conditions are determined by exile from East to West in the twenty-first century and the systems mentioned in Moser's definition can be identified as French and Iranian cultures. Instead of being directly transferred from one culture to another, as Moser's definition suggests, an encounter between different cultural values and ideas takes place within the narrative. In order for successful transfer to occur, it is argued that another element needs to be added, that of the reader who engages in the process. The implied reader of the original French version is Western and francophone, and the English translation of the text enables a wider audience to engage with the text. If the issue of language translation is significant for cultural transfer studies, this chapter focuses on the translation of cultural encounters rather than language translation in order to respect the theme of this volume, the association of cultural transfer and travel.

After the presentation of the network of exile and travel in relation to narrative, cultural encounters and transfer in the methodological framework, the first part of the analysis will focus on the metaphorical and physical levels of travel and exile as developed in the novel. The second part examines the narrator's confrontation with her sexual identity and the understanding of her hybrid body as a place of encounter between different cultures. The final part of the analysis discusses how the narrative creates moments of encounter between the narrator and her audience where the former assumes the role of cultural transmitter and meaningful cultural transfer can occur. In addition to offering new perspectives on the development of travel writing in the twenty-first century, this chapter contributes to cultural transfer studies by examining the role of the exiled narrator as cultural transmitter.

Exile, travel narrative and strange encounters[6]

The notions of travel and exile are commonly considered discrete categories and dealt with separately, although they are in fact strongly related through a series of categories characteristic of "contemporary figures of displacement".[7] Referring to Edward Said's work in which travel and exile exist alongside one another, Charles

---

6. This term refers to Sara Ahmed's book *Strange Encounters: Embodied Others in Post-Coloniality* (London: Routledge, 2000).

7. Charles Forsdick, "Travelling Theory, Exiled Theorists," in *Travel and Exile. Postcolonial Perspectives*, ed. Charles Forsdick (ASCALF Critical Studies in Postcolonial Literature and Culture 1, 2001), 1.

Forsdick demonstrates that these two concepts "coexist, conflict, intertwine, overlap,"[8] and that their relationship is ultimately contrapuntal. Kate Averis and Isabel Hollis-Touré equally underline this contrapuntal movement that identifies the experience of exile, given the continuous shifting between two different spaces, the old and the new, and the relationship between them. By examining existing categories of mobility, they propose to address and interrogate "the proximities and overlaps that exist between exiles, migrants, expatriates and travellers."[9] By leaving home, they all necessarily move away from the nation they grew up in, transgressing its ethnocentric environment. When considering "intersections and overlaps"[10] between different categories of movement rather than dealing with them separately, an apparent and joint effect is that of the undermining of nationalised categories. In the French literary field, such categorisation is clearly expressed in the distinction between Francophone and French authors for example, the first category usually associated with migration or exile literature, whereas the genre of travel is more likely to be attributed to the second category. An integral part of movement and travel however is the transgression of boundaries and categorisation, suggesting that a different figure, for example that of the nomad, translates the experience of mobility more effectively and extensively. In Gilles Deleuze and Félix Guattari's nomadology,[11] the nomad is presented as a stateless subject symbolising difference and transformation. Their nomad is a "machine de guerre" ("war machine") fighting against the state that represents fixation and classification. Nomad and state are always associated, since the one seeks to control the other, whereas this other has the desire to evade or destroy the state's control. In taking up Deleuze and Guattari's idea of difference, Rosi Braidotti has developed "a vision of feminist subjectivity in a nomadic mode"[12] that considers the female transcultural subject in its constant process of becoming. Becoming and difference are closely related: when becoming reaches the verge of stability, it differs into a new becoming, thus endlessly repeating its movement. This perspective corresponds to Said's idea of exile as a permanent state rather than a transient stage, since the condition of exile is always that of an outsider, it is a continuous state of "restlessness, movement, constantly being unsettled, and

8. Forsdick, "Travelling Theory, Exiled Theorists," 11.

9. Kate Averis & Isabel Hollis-Touré (eds.), *Exiles, Travellers and Vagabonds. Rethinking Mobility in Francophone Women's Writing* (Cardiff, University of Wales Press, 2016), 4.

10. Averis & Hollis-Touré, *Exiles, Travellers and Vagabonds*, 3.

11. Gilles Deleuze & Félix Guattari, "Traité de nomadologie: la machine de guerre," in *Mille Plateaux* (Paris: Minuit, 1980).

12. Rosi Braidotti, *Nomadic Subjects. Embodiment and Sexual Difference in Contemporary Feminist Theory* (New York: Columbia University Press, 1994), 1.

unsettling others."[13] In this respect, exile is not only an actual condition, but also a metaphoric one, comparable to the nomad as figurative concept in both Deleuze and Guattari's and Braidotti's theories, defined by "the subversion of conventions."[14] Despite these metaphorical and figurative levels, it has been argued that, unaware of their own privilege, postcolonial academics such as Said present "a narrative of class origin"[15] where exile is "an upper-class, émigré phenomenon,"[16] ultimately fetishizing the intellectual in exile. Sara Ahmed's criticism of Said demonstrates that scholarly works generally focus on exiles who find themselves in a privileged position, who have been able to study and continue their academic career in the new country for example. Criticisms of Braidotti's model are comparable to those of Ahmed and signal the risk of idealising and romanticising the nomad's rootlessness, since it discounts the negative effects of alienation, resulting for example in identity crises, feelings of nostalgia and homelessness. For Braidotti, "homelessness is a chosen condition,"[17] which, as Ahmed argues, is a privileged position, not taking into account subjects who are homeless "due to the contingency of 'external' circumstances."[18] This voluntary mobility generally neglects traumatic experiences and disorientation. The liberating and subversive experience of exile does however not exclude the equally occurring feelings of (identity) loss and non-belonging. They both shape the travelling subject whose transformation is achieved through the encounter of the liberating as well as the alienating experiences of exile. The result of this encounter is a transformation that particularly characterises all forms of displacement.

Considering mobility from this perspective of transformation and transgression, that of boundaries between the various categories of mobility, as well as the transgression of national and ethnocentric borderlines, this chapter adopts the perspective that travel as a mental and physical plot device is central to the process of defining and redefining the exile's identity, in particular in relation to space, culture and community. Reshaping individual identity includes a dynamic relationship between homeland and host country, between past and present. As Nacify has argued, exile culture is hybrid and involves "ambivalences, resistances, slippages, dissimulations, doubling and even subversion of the cultural codes of

13. Edward Said, "Intellectual Exile: Expatriates and Marginals," in *Representations of the Intellectual: the 1993 Reith Lectures* (New York, Pantheon Books, 1994), 53.

14. Braidotti, *Nomadic Subjects*, 5.

15. Ahmed, *Strange Encounters*, 210.

16. Ahmed, *Strange Encounters*, 210.

17. Braidotti, *Nomadic Subjects*, 17.

18. Ahmed, *Strange Encounters*, 83.

both the home and host societies,"[19] thus offering transnational perspectives and stories of cultural encounter.

The experience of exile is translated into text, permitting "contestation of discourses of assimilation, citizenship, nationalism, and diaspora"[20] and often undermining stereotypes and prejudices. These narratives also offer the possibility of translating cultural encounters into meaningful transmissions of ideas, values, beliefs and practices by creating contact zones[21] where different experiences meet as well as clash. If these zones take diverse shapes, from geographical spaces (e.g. Brandau in this volume) to chronotopical entanglements (e.g. Sandrock in this volume) and mystical dimensions (e.g. Yuan in this volume), they always allow for cultural encounters to become productive events of fusion as well as conflict. Cross-cultural translation, collision and blending of received cultural codes lay the foundation for successful occurrences of cultural transfer. When focusing on women's experiences in particular, the dialogic between sameness and difference often corresponds to an exploration of the self in relation to others. As studies of gender and travel narratives have demonstrated,[22] women writers have built on the genre's potential to give access to unfamiliar cultures and to wield literary and cultural authority. Petra Broomans' chapter in this volume explains how Mary Wollstonecraft constructs a new form of travel writing from a gendered perspective. This perspective is further developed in women's migrant writing, where the female displaced subject offers encounters with different cultures while telling her own story and including others in it. Taking up life writing as part of the travel narrative permits an investigation of the feeling of alterity within the self and offers "a way of overcoming the impasse between self and others."[23] In the process of (re)constructing subjectivity and alterity, "received codes of gender, sexuality, family and the body"[24] are cross-culturally transmitted. As Petra Broomans has argued, "cultural transmitters are actors in a transnational space, and ideas and

19. Hamid Nacify, *Making of Exile Culture: Iranian Television in Los Angeles* (Minneapolis: University of Minnesota Press, 1993), xvi.

20. Eva C. Karpinski, *Borrowed Tongues: Life Writing, Migration, and Translation* (Waterloo (Ont.), Wilfrid Laurier University Press, 2012), 5.

21. Mary Louise Pratt has defined the term contact zones as "social spaces where different cultures meet, clash, and grapple with each other, often in highly asymmetrical relations of domination and subordination – like colonialism, slavery, or their aftermaths as they are lived out across the globe today." (*Imperial Eyes: Travel Writing and Transculturation*, London: Routledge, 1992, 4).

22. See for example Elizabeth A. Bohls' *Women Travel Writers, Landscape and the Language of Aesthetics, 1716–1818* (Cambridge: Cambridge University Press, 1995).

23. Karpinski, *Borrowed Tongues*, 29.

24. Karpinski, *Borrowed Tongues*, 44.

images travel through time and across borders."[25] The first category of cultural transmitters as distinguished by Paul Van Tieghem that Broomans refers to, "les individus (individuals),"[26] could also include writers who cross national and linguistic boundaries and connect the different realities of home and host countries, for example those regarding gender and the body. All encounters with the other, or strange encounters in the words of Ahmed, "are played out on the body."[27] Difference, Ahmed argues, "is established as a relation between bodies."[28] Travel writing, drawing on the expression of the self in life writing, redefines the displaced body in relation to bodily others, thus creating the possibility of meaningful encounters and transfers.

This framework has presented a reflection on the notions that are key for the analysis of Djavadi's novel, a narrative in which the exiled female protagonist recounts her travels by exploring her disoriented self through the encounter with others. This exploration is characterised by the sense that her body is out of place, in geographical, social as well as gender terms, but also by the belief that cultural and bodily connection(s) will foster a re-orientation of her identity.

Real and narrated forms of travel

Kimiâ's story of exile starts in 1981 when, at the age of ten, she flees Iran with her mother and two sisters to join their father in France.[29] Her experience can be compared to many Iranians fleeing the country after the establishment of an Islamic Republic in 1979, under the supreme leader Khomeini. Many intellectuals, including Kimiâ's father, went into exile for political reasons, followed by their families. During the 1980s, when the revolution developed into a reign of terror and the armed conflict between Iran and Iraq had devastating effects on the country's economy, pushing it into a deep crisis, the migration flux further increased.

25. Petra Broomans, "The Meta-Literary History of Cultural Transmitters and Forgotten Scholars in the Midst of Transnational Literary History," in *Cultural Transfer Reconsidered. Transnational Perspectives, Translation Processes, Scandinavian and Postcolonial Challenges*, eds. by Steen Bille Jørgensen & Hans-Jürgen Lüsebrink (Leiden: Brill/Rodopi, 2021), 79.

26. Broomans, "The Meta-Literary History of Cultural Transmitters and Forgotten Scholars in the Midst of Transnational Literary History," 81.

27. Ahmed, *Strange Encounters*, 39.

28. Ahmed, *Strange Encounters*, 44.

29. In France, the Iranian diaspora has produced a prolific literature by women writers such as Chahdortt Djavann, Marjane Satrapi, Maryam Madjidi or Abnousse Shalmani, interrogating and redefining notions of exile, home, language, family and history. They either arrived in France at a young age with their families, like Négar Djavadi, or as young adults.

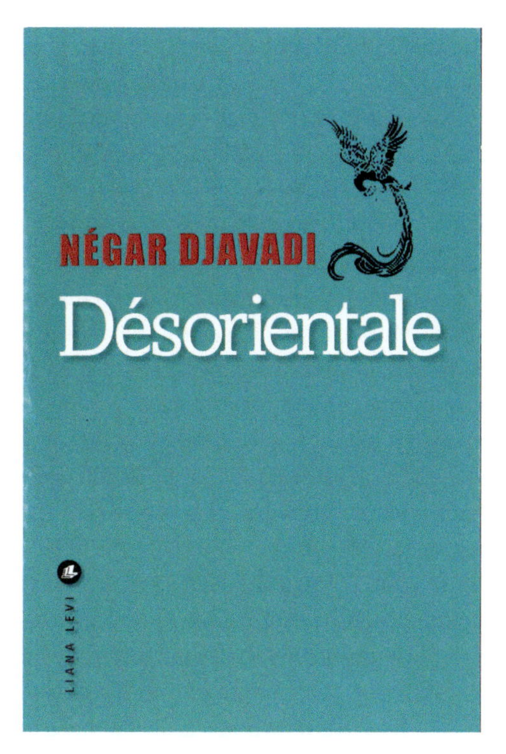

@https://www.lianalevi.fr/catalogue/desorientale/

In 1985, the number of Iranian immigrants living in France reached the number of 22.484, four times as many as in 1979.[30] As Vida Nassehi-Behnam has argued in her article on the Iranian diaspora in France, a large number of women were able to secure their place in French society by becoming economically active, participating in language courses and by joining the local school's parent committee for example.[31] Kimiâ's mother Sara had been introduced to French culture as a student after having received a scholarship to study in France for a year in order to write her thesis on Jean-Jacques Rousseau. The narrator refers to this period as "a mirage in the middle of the desert,"[32] followed by her mother's "immoderate passion for France."[33] Since also her father, Darius, studied in France where

30. Nader Vahabi, "La diaspora iranienne en France. Profil démographique et socioéconomique," *Revue Migrations et Société* 158 (2015): 37.

31. Vida Nassehi-Behnam, "Diaspora iranienne en France: changement et continuité." *CEMOTI, Cahiers d'Études sur la Méditerranée Orientale et le Monde Turco-Iranien* 30 (2015): 141.

32. Djavadi, *Disoriental*, 41.

33. Djavadi, *Disoriental*, 41.

he obtained his doctorate in philosophy at the Sorbonne, French culture and language have been an integral part of the Sadr family's life in Tehran. Sara enrolled her daughters at the Lycée Razi, the French school "located in a residential area in north Tehran"[34] and Darius was nicknamed "*Amou Farançavi*, French Uncle"[35] by the rest of the family.

Unlike her sisters, Kimiâ does not like the French language as a child, mostly because her fellow students consider it to be "superior to Persian"[36] and feel that they are themselves "superior to other Iranians, to the teeming uncultivated masses lost in the depths of the Middle East"[37] because they speak French. As a young girl, she does not share her mother's passion for France and her critical attitude develops into aversion during her first experience on French territory, the French embassy in Istanbul. After having been smuggled out of Iran, having passed through cities and villages, having crossed the mountains of Kurdistan on horseback and by foot, Sara and the girls expect to be welcomed "as stateless Francophiles"[38] once they have entered the embassy. The feeling of being "on the threshold of the promised land"[39] turns into a traumatic experience when they are treated in a condescending manner and dismissed because the ransom they had paid to their smugglers did not include their visa, and they have no money left to pay for it. Particularly the fact that the officials use French slang words in order to confuse them seems to hurt Kimiâ who feels as if a knife has been deliberately planted in their backs and realises that they have "no allies, no friends, no refuge"[40] and nowhere to go. The 10 year old senses that the contempt with which they are treated is a precursor to the situation they will face in exile. When they finally receive the necessary documents and are allowed to leave for the alleged promised land of Rousseau, Hugo and Sartre, the girl feels only anger and mistrust.

The promised land is indeed quite different from the idealised picture that particularly Sara had always presented to her daughters. The narrator reflects on the alienating experience of exile as a second birth: "Soon, I will be born for the second time. Accustomed to coming into the world amidst blood and confusion, to awakening Death and inviting it to the party, this rebirth [...] is undeniably worthy of the first."[41] The use of the words blood, confusion and Death

34. Djavadi, *Disoriental*, 41.
35. Djavadi, *Disoriental*, 55.
36. Djavadi, *Disoriental*, 42.
37. Djavadi, *Disoriental*, 42.
38. Djavadi, *Disoriental*, 242.
39. Djavadi, *Disoriental*, 242.
40. Djavadi, *Disoriental*, 244.
41. Djavadi, *Disoriental*, 248.

stresses the violence associated with the exile's existence. The first moment of estrangement is related to her first name which "won't be pronounced in the same way anymore; the final *â* will become *a* in Western mouths, falling silent forever."[42] This different pronunciation of her name, marking a period of change and silence, undeniably estranges her from her oriental roots, without bringing her closer to Western habits: "Soon, I will be disoriental"."[43] This feeling of being disoriented is strongly related to the unfamiliarity of the language: "there were so many words and names that I just didn't have."[44] This lack of words creates a continuous distance between the young girl and the world around her, since she doesn't know how to talk about it, turning the "talkative sociable child"[45] that she used to be into a "wordless girl."[46] The impossibility to express herself feels like an injury, a wound that barely heals and becomes a "scar that runs across [her] vocabulary."[47] When the words do come out, she is constantly reminded of the fact that she has an accent, a comment immediately followed by the question: "where are you from?"[48] If this question appears to be innocent, and might even suggest genuine interest, its reiterative nature is a painful reminder of difference: "no one misses the foreigners. No one can resist the cheap pleasure of scratching that itch of difference."[49] Commenting on language use, whether it be grammatical errors, vocabulary or pronunciation, is an obvious means of stressing the other's strangeness and non-belonging. In order to avoid further scratching, there is a retreat into silence. Despite the feeling that her voice often lodges in her "throat as if in a tomb,"[50] Kimiâ does not distance herself further from the language, but rather transforms her muteness into a metaphorical travel by creating a narrative in which she establishes a connection between the past and the present, here and there, French and Iranian culture. Storytelling develops into a means of overcoming the traumatic experience of exile, which is reflected in the narrative structure of the text. Stories referring to her first years in exile, to her childhood in Tehran, to her family saga or the history of Iran, seamlessly flow into comments on her current situation which is that of the waiting room in a Parisian fertility clinic, or reflections on more recent events in her life including

42. Djavadi, *Disoriental*, 248.
43. Djavadi, *Disoriental*, 248.
44. Djavadi, *Disoriental*, 111.
45. Djavadi, *Disoriental*, 54.
46. Djavadi, *Disoriental*, 111.
47. Djavadi, *Disoriental*, 112.
48. Djavadi, *Disoriental*, 117.
49. Djavadi, *Disoriental*, 117.
50. Djavadi, *Disoriental*, 117.

friendships and relationships, for example with her two sisters. The narrative of her arrival in Paris and her second birth for example is followed by this observation: "I turn. A man in a wheelchair is glaring at me as if he's been waiting here for ten years. That's Paris, too – that look. It reflects what this city does to people, whether they're natives or from anywhere on the planet."[51] The narrator refers to the front door of the clinical laboratory which she is about to exit with the result of her test when another patient is eager to pass and leave the building before she does. She recognizes the look in his eyes reflecting impatience, a feeling that does not discriminate between the patients' origins but seems identical to everyone visiting a hospital. The fictional account of her parents' love story resulting in their wedding ceremony in the 1960s in Tehran smoothly blends into the narrator's appointment taking place decades later at the sperm bank on the Boulevard du Port-Royal in Paris. These examples illustrate that her personal story is part of a bigger picture; she is connected to a history that shaped her parents and their relationship, as well as to similar experiences lived by people she observes but does not develop a relationship with.

Storytelling enables the narrator to go back and forth between various moments in her life, comparing and contrasting manners and routines. The everyday reality of French society and her outsider status on the contrary remain difficult to comprehend. Where they seem to connect on a narrative level, encounters in real life are mainly characterised by conflict. Not only her accent and grammatical errors, also her Iranian origin is an incentive for painful moments: "the French had a disastrous image of Iran [...], Iran had acquired an unshakeable reputation as a fanatical backwater at war with the West."[52] This image is often projected onto her: "Iran was [...] an imaginary country full of Muslim fundamentalists of whom I suddenly became the representative."[53] During her adolescent years, Kimiâ increasingly starts to avoid this painful reality by looking for a place "where Iran, and France [don't] exist,"[54] and immerses herself in the Parisian underground scene, "a meeting place for teenagers estranged from their families"[55] where no one asks where she's from nor corrects any grammatical errors. She goes into hiding, away from her family and her roots, and from what she has come to see as her "foreign face."[56]

51. Djavadi, *Disoriental*, 249.

52. Djavadi, *Disoriental*, 262.

53. Djavadi, *Disoriental*, 117.

54. Djavadi, *Disoriental*, 265.

55. Djavadi, *Disoriental*, 267.

56. Djavadi, *Disoriental*, 277.

To travel then becomes a way of hiding, a habit that she seems to have inherited from her father, who disappeared from time to time from the country and would, for the rest of his life, "hold onto the habit of running away."[57] The narrator notes that "part of him had always been an exile,"[58] thus inadvertently explaining her own state of being. Exile and travel are intertwined and a necessary part of her father's life as much as they are part of hers. From Paris she travels to Brussels at the age of eighteen, moving on to Berlin, and then to Amsterdam and London, often working in bars where she can "stay hidden in the dark."[59] Her long hair that offers her a means to hide her timidity is like a curtain behind which she can "peep[...] out at the world."[60] This symbolises not only her reluctance to expose herself to the world, but also her efforts to "gain some self-assurance"[61] and, a fortiori, a better understanding of her feeling of being disoriental. Through a series of strange encounters during her travels she understands that she cannot spend her life hiding, but has to address her disoriented state by turning to her body.

The hybrid body and gender norms

Acknowledging and understanding her body is part of a long process that started in her early childhood. Before attending school, Kimiâ did not consider her body being different from that of her sisters for example. When preparing for her first day of school, she realises for the first time that she has never considered her body being that of a girl. After being dressed in the new clothes her mother bought for the occasion, she looks at herself in the mirror:

> What I felt then was so strange and unexpected that it paralyzed me: the sense of being dressed up as a girl, and the sudden awareness that I *was* one. A feeling that was quickly joined by panic at the thought of having to go to school dressed like this, and behaving accordingly. How was I going to manage it?[62]

The young girl quickly decides to hide her panic at her mother's and sister's euphoric reactions and follows her sisters on the school bus. The mechanism of hiding will determine her behaviour until the moment she accepts her lesbian sex-

57. Djavadi, *Disoriental*, 79.
58. Djavadi, *Disoriental*, 79.
59. Djavadi, *Disoriental*, 283.
60. Djavadi, *Disoriental*, 277.
61. Djavadi, *Disoriental*, 277.
62. Djavadi, *Disoriental*, 172.

ual identity, which only occurs at the end of a long process of denial, confusion, and finally acknowledgment.

Retrospectively, the fact that she differs from her sisters was firstly acknowledged by her grandmother Emma, who had the habit of predicting the gender of unborn children by reading the pregnant woman's coffee grounds, and declared that this child would be a boy. When, after the child's birth, she visits her daughter and granddaughter instead of grandson at the hospital, Emma's thoughts beautifully illustrate gender performance norms: "How would she muddle through life, with a body that had been tampered in such a way?"[63] Since Emma is convinced the girl's body is supposed to be a boy's, she predicts that accepting her body will be problematic for this child. When reacting to her daughter's triumphant "Did you see? It's a girl…"[64] by saying "Not quite,"[65] Emma acknowledges the possibility of a fluid gender identity that is ignored in Iranian society, as the narrator explains further in the novel: "The term "tomboy" doesn't exist; nor does any other term or word that recognizes that difference. You're a boy or a girl, and that's that."[66] This is a perfect example of gender performativity as a ritualised socially constructed norm produced by the "reiterative power of discourse."[67] As the narrator further explains: "Down through the generations, codes have been put in place. Certain codes for raising boys, and others for raising girls. […] It's about the future. About becoming a husband and father […]. Or becoming a wife and mother […]. No one knows how to raise the in-betweens, or deal with the not-quites."[68] This is illustrated by the fact that she is also supposed to fall into either of the categories; her mother raises her the same way she raises her sisters, her father treats her like "his imaginary son"[69] and cuts her hair as short as a boy's. Only Emma seems to notice her difference but is only able to explicitly voice it in a letter addressed to her daughter ten years after their departure for Paris: "Everything you were afraid of when you thought you were carrying a boy, she will do – as she has always done. Even, and this is the hardest part for me to write, bring you a daughter-in-law. […] if your daughter prefers girls, let her prefer girls."[70]

---

63. Djavadi, *Disoriental*, 146.

64. Djavadi, *Disoriental*, 146.

65. Djavadi, *Disoriental*, 146.

66. Djavadi, *Disoriental*, 211.

67. Judith Butler, *Bodies That Matter. On the Discursive Limits of "Sex".* (London/New York: Routledge, 1993).

68. Djavadi, *Disoriental*, 211.

69. Djavadi, *Disoriental*, 161.

70. Djavadi, *Disoriental*, 287.

Kimiâ herself is confronted with her sexual identity at the age of ten in a very brutal and painful way. Preferring "to play ball with the boys,"[71] she initially sees no harm in the fact that she has little in common with other girls, until the day her sister Leïli whispers a word in her ear that turns her world upside down.[72] Her sister murmurs it in French so that no one else will understand because it causes too much shame: lesbienne. At that moment Kimiâ realises she is excluded by the other girls; the word means she has "become something monstrous."[73] Her first reaction to this exclusion from the cultural norm is to appropriate it no matter what it takes, to "change into a new form" in order to "turn into a girl like the other girls."[74] She desperately wants to fit in, but is pushed into an uninhabitable zone since others continuously exclude her as an abject[75] being. As observed earlier, Kimiâ reacts to this unliveable zone of social life by hiding herself and running away. During this nomadic period she meets her future partner Anna, who is "the complete opposite side of the coin"[76] from the narrator, and a mystery to her. In this relationship, in which she mirrors herself in her opposite other, she finally becomes a subject circumscribing her "own claim to autonomy and to life."[77] By coming to terms with her lesbianism, she is able to face the truth: "I'd finally managed to strip myself of everything; I wasn't anything but myself anymore."[78] The epitome of her claim to autonomy is her pregnancy; the couple hopes to have a baby through artificial insemination, and, after a year and seven months, their desire is fulfilled. As a lesbian and a future mother, she has defied the gender codes with which she was raised. She has broken down prejudices by demonstrating that she is "just an ordinary girl,"[79] despite the resistance she encountered.

When walking out of the Parisian hospital after having received the news, Kimiâ's pregnant body functions as a contact zone, incorporating her Iranian origins, and at the same time connecting with her Flemish partner Anna, their French sperm donor Pierre, and the Parisian environment where they will raise their twin babies. As a child, being ashamed of the strangeness of her body and understanding that others do not consider her a girl, she dreamed of having a

71. Djavadi, *Disoriental*, 162.

72. Djavadi, *Disoriental*, 206.

73. Djavadi, *Disoriental*, 209.

74. Djavadi, *Disoriental*, 201.

75. As Judith Butler notes, "the notion of *abjection* designates a degraded or cast out status within the terms of sociality", *Bodies That Matter*, Note 2, 243.

76. Djavadi, *Disoriental*, 289.

77. Butler, *Bodies That Matter*, 3.

78. Djavadi, *Disoriental*, 291.

79. Djavadi, *Disoriental*, 120.

body like the "lead singer of Bauhaus"[80] and of shaping it the way she wanted it. After having experienced alienation and exclusion since her childhood, she is able to acknowledge her feelings of disorientation by accepting her "hybrid body."[81]

Exile, as a second birth when arriving in Paris, makes her go through the whole process of disorientation again. She disappears inside her body when she is confronted with a strange world she doesn't "know how to talk about,"[82] then goes into hiding and runs away, since she is unable to "inhabit fully the present or present space."[83] Her temporary hiding place of the Parisian underground seems to offer a first possibility to "exist in an infinite *now*."[84] Here she learns that "being homosexual or heterosexual doesn't mean anything"[85] and this nightly Paris presents itself as a "liberating and insidious place."[86] The strange encounters that occur during this underground period do however not result in real contacts or connections; the protagonist always keeps her distance from equally estranged others and ultimately wants to "dissolve and disappear."[87] As Ahmed argues, migrant bodies inhabit a space that is "neither within nor outside bodily space"[88] and Kimiâ's dislocation of migration seems to reach its culminating point when she becomes totally indifferent towards her self-destructive way of life. As Ahmed explains, the void or absence resulting from migration and bodily estrangement can be reinhabited "through gestures of friendship with others who are already recognised as strangers."[89] In Kimiâ's case, not the encounter with other migrants nor with outsiders from the underground scene, but her acquaintance with another stranger to the heterosexual norm enables her to reinhabit her disoriented body. This relationship that develops with intermissions and is confirmed by the desire to raise a child, results in the narrator's acknowledgement not only of what she once thought was her "bizarre physique,"[90] but also of her exiled body. The acceptance of her body as her "only country"[91] thus coincides with the acknowledgement of her disorientation and exile.

80. Djavadi, *Disoriental*, 267.
81. Djavadi, *Disoriental*, 267.
82. Djavadi, *Disoriental*, 111.
83. Ahmed, *Strange Encounters*, 92.
84. Djavadi, *Disoriental*, 268, italics in the text.
85. Djavadi, *Disoriental*, 268.
86. Djavadi, *Disoriental*, 267.
87. Djavadi, *Disoriental*, 273.
88. Ahmed, *Strange Encounters*, 92.
89. Ahmed, *Strange Encounters*, 93.
90. Djavadi, *Disoriental*, 268.
91. Djavadi, *Disoriental*, 267.

Having regained the self-assurance she experienced as a child before being stigmatised as different, she becomes someone who translates herself "into other cultural codes"[92] in order to forge a future. Her pregnant body represents the encounter between cultures and contains past, present, and future, which she is willing and capable to share with others.

(Meta)narrative, contact zones and cultural transfer

As observed in the first part of this analysis, storytelling is essential in the process of overcoming the traumatic experience of exile. By narrating her personal experience in relation to her family members and to the people she meets in strange encounters during her travels, the protagonist establishes a new relationship with her disoriented self. The painful impact of this disorientation can be transformed into productive effects by means of storytelling. This productivity translates into the use of two narrative techniques creating a connection with the audience that will be examined in this final part of the analysis: firstly, the narrator addresses the implied reader directly throughout the narration, and secondly she regularly reflects on her own text, thus establishing a metanarrative that includes the audience in her considerations. As will be demonstrated in this section, these techniques create contact zones in which the narrator assumes her role as cultural transmitter.

The form in which she presents her narrative blends Iranian and Western traditions, since, for Iranians, telling stories is "one way to deal with a fate consisting of invasions and totalitarianism"[93] as, according to the narrator, "telling and retelling, embellishing, and lying, [...] means staying alive."[94] The use of the first person perspective however is part of the Western literary tradition since, as Goldin has argued, "throughout most of Iranian history, modesty and secrecy prevented Iranian women from recording their life narratives."[95] The genre has become widely popular among Iranian women writers in exile writing in the adopted language,[96]

92. Djavadi, *Disoriental*, 54.

93. Djavadi, *Disoriental*, 54.

94. Djavadi, *Disoriental*, 54.

95. F.D. Goldin, "Overcoming gender: The impact of the Persian language on Iranian Women's Confessional Literature," in *Familiar and Foreign: Identity in Iranian Film and Literature*, eds. Manijeh Mannani & Veronica Thompson (Alberta: Athabasca University, 2015), 40.

96. See for example Azar Nafisi, *Reading Lolita in Tehran: A Memoir in Books*. New York: Random House, 2003; Azadeh Moaveni, *Lipstick Jihad: A Memoir of Growing Up Iranian in Amer-*

reflecting "their comfort with English and French as languages that are more suitable for carrying their voices, languages not delimited by Iranian cultural boundaries."[97] Djavadi's book thus complies with a recent trend in Iranian women's writing in exile, although the majority of the works are categorised as memoir writing, whereas the first person perspective in *Désorientale* is principally related to travel narrative, as discussed in the theoretical framework. The amalgamation of the two traditions constitutes the foundation for the narrator's role as cultural transmitter; a distinct example of this is the transmission of the story of Kimiâ's mother as part of the travel narrative. During her exile, Sara has written her memoir, *Our Life*, "describing the Revolution years up to [their] departure for France."[98] She wrote it in Persian, and since it "became a bestseller in Tehran",[99] she would have loved one of her daughters to translate it into French. Kimiâ, nor her sisters, translates their mother's book, but the narrator inserts the pages that their mother translated herself in *Désorientale* and reflects on it. This reflection reveals her own urge to write: "she had tried to create some order out of the chaos of her experiences and feelings, the fiery tide of the past that was consuming her from the inside. This wasn't only an account for posterity, it was an urgent need."[100] The narrator clearly recognizes this need and instead of translating her mother's text directly, she imagines how Sara would have finished the story where she has left off. The part Sara translated into French recounts their flight from Iran, crossing the mountains of Kurdistan, and the anxiety she experienced during this difficult journey. It finishes in the middle of sentence when they have reached Turkey. Her daughter continues the story of their exile in the third person, thus confronting her mother's story with her own. This ultimately leads up to the account of a conflict between mother and daughter during adolescence, followed by the latter taking refuge in the Parisian underground scene. The text functions as a contact zone where their experiences meet as well as clash; the narrator is the mediator of her mother's story as well as the witness of the incomprehension between them.

In this constant movement between the need to tell her own story and mediating the stories of others, she establishes a pact with the reader: "you'll have to be patient with me, Dear Reader",[101] since she is committed to disclose even the most difficult events that shaped her life, but she needs to prepare for them. The audi-

ica and American in Iran. New York: Public Affairs, 2005; Sorour Kasmaï, *La vallée des aigles. Autobiographie d'une fuite.* Paris: Actes Sud, 2006.

97. F.D. Goldin, "Overcoming gender", 39.

98. Djavadi, *Disoriental*, 167.

99. Djavadi, *Disoriental*, 235.

100. Djavadi, *Disoriental*, 237.

101. Djavadi, *Disoriental*, 165.

ence engages in the process and find themselves involved from the start: "You, the citizens of this country, with your income taxes and compulsory deductions and council taxes – but also your education, your intransigence, your critical minds and your spirit of solidarity and pride and culture and patriotism, your devotion to the Republic and democracy [...]."[102] The narrator is clearly addressing a French audience, whom she invites to reflect on some of their cultural values. She also announces her project: "Talking about the present means I have to go deep into the past, to cross borders and scale mountains and go back to that lake so enormous they call it a sea"[103] and warns the reader that her account will be non-linear, that her memory is imperfect and her story an entanglement of various threads. While guiding the reader through these threads, the narrator establishes a relationship where she asks her audience for patience and understanding. In a tongue-in-cheek way she refers to the historical and geographical contexts the reader needs to be aware of, for example in footnotes: "To make things easier for you and save you the trouble of looking it up on Wikipedia, here are a few facts"[104] or: "I'll just remind you here that an attempt to assassinate Mohammed Reza Pahlavi took place on February 4, 1949."[105] She also anticipates their reaction by stating "I know what you are thinking"[106] or "I'm sure you would prefer"[107] and comments on the time/space construction of her text: "End of flash-forward";[108] "so now we're back in Egypt again";[109] "let me take you back in time again."[110]

These asides seem to invite the audience to a critical reflection not only on the content and structure of her narration, but also on their own values and prejudices. Have they not, themselves, given in to "the cheap pleasure of scratching that itch of difference"[111] when meeting a foreigner, like many of the protagonist's interlocutors? And do they not think "the girl whose father wanted a son acts like a boy and ends up a lesbian, what a cliché?"[112] The narrator admits this is true, but only "when you have access to books, and movie theatres where they show *Sylvia Scarlett* or *The Bitter Tears of Petra von Kant.* When you've absorbed May

102. Djavadi, *Disoriental*, 11.
103. Djavadi, *Disoriental*, 12.
104. Djavadi, *Disoriental*, 21, Note 1.
105. Djavadi, *Disoriental*, 77, Note 6.
106. Djavadi, *Disoriental*, 69; 211.
107. Djavadi, *Disoriental*, 216.
108. Djavadi, *Disoriental*, 103.
109. Djavadi, *Disoriental*, 67.
110. Djavadi, *Disoriental*, 305.
111. Djavadi, *Disoriental*, 117.
112. Djavadi, *Disoriental*, 211.

’68 and the Sexual Revolution, and the feminist movement and Simone de Beauvoir. When you’ve listened to The Runaways and Bowie and Patti Smith.”[113] In Tehran where she spent the first decade of her life, there was a different reality with clearly distinguished gender roles. The narrator also confronts the audience with their ignorance of today’s Iran, where “transsexuality exists because there is something worse than being transsexual, and that’s being homosexual.”[114] In the Islamic Republic, homosexuality is “a crime punishable by death. Women as well as men, sometimes only teenagers, are blindfolded and hanged from cranes in public.”[115] This dreadful image emphasises that sexual freedom is not self-evident. As a lesbian, she could risk her life when returning to Iran. A reunion with her country of origin is therefore not plausible, and is reminiscent of her father’s situation who, as a journalist writing critically about the regimes of both the Shah and the Ayatollah, never returned to Iran. He continued writing in Paris, using his pen as a weapon to denounce oppression, and eventually paid for it with his life. In their Parisian apartment, he is assassinated by agents of the Iranian secret police, a homicide that the narrator throughout the text refers to as THE EVENT, the account of which she defers as long as possible. It is one of the most painful events in her life, illustrating the necessity and the danger of putting pen to paper and expressing one’s ideas. Telling stories, writing them down, is a survival mechanism as much as it is a highly hazardous act. When creating a narrative is a means to reinvent the self, it is also a form of exposure. In telling her story, she puts herself at risk, hoping to establish a connection with others.

In her narrative, the narrator scrapes the surface of dissimilarities between the two cultures, and at the same time creates moments of contact and understanding. By openly sharing her experiences and sometimes yielding to the temptation to give “a crash course on modern Iranian history,”[116] she paints the image of personal experiences in connection to divergent traditions and cultural values. By reflecting on the fact that “to really integrate into a culture, [...] you have to *dis*integrate first, at least partially, from your own. You have to separate, detach, disassociate,”[117] she invites her audience to participate in her own disorienting experience, to open up to the other and to defer judgement.

113. Djavadi, *Disoriental*, 211.

114. Djavadi, *Disoriental*, 212.

115. Djavadi, *Disoriental*, 315.

116. Djavadi, *Disoriental*, 263.

117. Djavadi, *Disoriental*, 112.

Conclusion

This chapter has examined the intersection of the concepts of travel and exile through a number of conflictual and intertwining movements in *Désorientale.* As observed in the theoretical framework, travel and exile both collide and overlap. The emphasis, proposed for example by Averis and Hollis-Touré, on intersections between different categories of movement rather than separating them, allows for a discussion of mobility from the perspective of transformation and transgression that also takes into account the pain of displacement. Rather than idealising uprootedness, as scholars such as Said and Braidotti are criticised for, this chapter has insisted on the conjunction between liberating and alienating experiences of the travelling subject. This concurrence particularly takes place in contact zones where received cultural codes collide as well as meet, and which lay the foundation for successful occurrences of cultural transfer. Travel narratives offer the possibility to translate such cultural encounters into meaningful transmissions of ideas and emotions, since they allow for a redefinition of the exiled self in relation to others. Following Moser's categorisation, the travel narrative thus functions as the material support that enables the subject, the narrator, to transfer immaterial objects. These are, on a personal level, ideas and emotions, and, with regard to culture, values, practices and codes. By studying the travel narrative as a mental and physical plot device, this chapter has analysed the narrator's body and narrative as zones of contact and exchange where, in Ahmed's words, strange encounters occur.

In *Désorientale,* these encounters are cultural as well as individual, alienating as well as liberating, and allow for the exiled narrator to establish a new relationship with her disoriented self. For Kimiâ, the encounter with French language and culture is a disorienting experience which is inevitably connected with her gender and sexual disorientation that deviates from the heterosexual norm. In order to escape the painful feelings of cultural and bodily estrangement, she goes into hiding and prefers to be silent. Her nomadic life illustrates her restlessness and otherness, but also results in the creation of a narrative where the different threads of time and space are connected and where the self can be reinvented. Language plays a pivotal role in the confrontation with the self, and silence turns into narrative when the strangeness of the hybrid body is no longer a discomfort. Ultimately, the body as well as the narrative function as contact zones where experiences can be shared with others. While establishing contact with the audience, the narrator critically assesses personal as well as cultural values and ideas and invites the reader to share her experience of (dis)integration and reflect on received cultural codes of family, gender, sexuality and the body in relation to homelessness and belonging. In the process of transfer that is developed by means of the travel narrative, personal and cultural encounters allow for the narrator to transmit emotions, ideas and values to Western audiences.

References

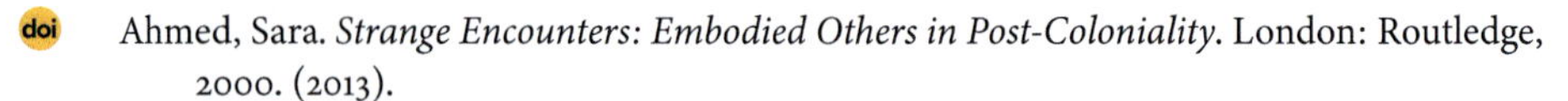

Ahmed, Sara. *Strange Encounters: Embodied Others in Post-Coloniality*. London: Routledge, 2000. (2013).

Averis, Kate and Isabel Hollis-Touré (eds.). *Exiles, Travellers and Vagabonds. Rethinking Mobility in Francophone Women's Writing*. Cardiff: University of Wales Press, 2016.

Braidotti, Rosi. *Nomadic Subjects. Embodiment and Sexual Difference in Contemporary Feminist Theory*. New York: Columbia University Press, 1994.

Broomans, Petra. "The Meta-Literary History of Cultural Transmitters and Forgotten Scholars in the Midst of Transnational Literary History." In *Cultural Transfer Reconsidered. Transnational Perspectives, Translation Processes, Scandinavian and Postcolonial Challenges*, edited by Steen Bille Jørgensen & Hans-Jürgen Lüsebrink, 64–87. Leiden: Brill/Rodopi, 2021.

Butler, Judith. *Bodies That Matter. On the Discursive Limits of "Sex"*. London/New York: Routledge, 1993.

Deleuze, Gilles et Félix Guattari. "Traité de nomadologie: la machine de guerre." In *Mille Plateaux*. Paris: Minuit, 1980, 434–527.

Djavadi, Négar. *Disoriental*. Translated by Tina Cover. New York: Europa Editions, 2019.

Forsdick, Charles. "Travelling Theory, Exiled Theorists." In *Travel and Exile. Postcolonial Perspectives*, edited by Charles Forsdick, 1–13. *ASCALF Critical Studies in Postcolonial Literature and Culture* 1, 2001.

Goldin, F. D. "Overcoming gender: The impact of the Persian language on Iranian Women's Confessional Literature." In *Familiar and Foreign: Identity in Iranian Film and Literature*, edited by Manijeh Mannani & Veronica Thompson, 31–60. Alberta: Athabasca University, 2015.

Greenblatt, Stephen. "A Mobility Studies Manifesto." In *Cultural Mobility: A Manifesto*, edited by Stephen Greenblatt, with Ines G. Županov, Reinhard Meyer-Kalkus, Heike Paul, Pál Nyíri, & Friederike Pannewick, 250–253. Cambridge, Cambridge University Press, 2010.

Karpinski, Eva C. *Borrowed Tongues: Life Writing, Migration, and Translation*. Waterloo (Ont.): Wilfrid Laurier University Press, 2012.

Moser, Walter. "Pour une grammaire du concept de « transfert » appliqué au culturel." In *Transfert: Exploration d'un champ conceptuel*, edited by Pascal Gin, Nicolas Goyer, & Walter Moser, 49–75. Ottawa: Les Presses de l'Université d'Ottawa, 2014.

Naficy, Hamid. *Making of Exile Culture: Iranian Television in Los Angeles*. Minneapolis: University of Minnesota Press, 1993.

Nassehi-Behnam, Vida. "Diaspora iranienne en France: changement et continuité." *CEMOTI, Cahiers d'Études sur la Méditerranée Orientale et le Monde Turco-Iranien* 30 (2015): 135–150. https://www.persee.fr/doc/cemot_0764-9878_2000_num_30_1_1555. Accessed January 31, 2021.

Said, Edward. "Intellectual Exile: Expatriates and Marginals." In *Representations of the Intellectual: the 1993 Reith Lectures*, 47–64. New York: Pantheon Books, 1994.

Vahabi, Nader. "La diaspora iranienne en France. Profil démographique et socioéconomique." *Revue Migrations et Société* 158 (2015): 19–40.

Index